"It is remarkable that any electron flow through the cellular gels is largely ignored in classical texts on cell physiology, which also incorrectly treat the intracellular milieu as a fluid. Dr Noble offers a novelty and tantalizing challenge to the aspiring physiologist by suggesting the incorporation of electron distribution on the cellular gel proteins in the physiology of electrical and contractile properties. The book, by doing so, brings cell physiology indeed closer to the realm of modern physics."

Henk E D J ter Keurs
Professor Emeritus, University of Calgary, Canada

"Mark Noble has had a long and distinguished career in academic and cardiac medicine ... much of which has focussed on electrophysiological mechanisms controlling rate, rhythm and contractile behaviour in the heart. This book, which draws on reflections in his retirement, is a fascinating and iconoclastic survey of electrophysiological processes in both cardiac and many other excitable tissues in the animal kingdom which will interest and alter the understanding of both clinicians and basic scientists in many disciplines."

Anthony Seed
Professor Emeritus, Imperial College London, UK

"This book delivers a critical analysis of electric phenomena in physiology from the cellular to the systemic level. Dr Noble's own very successful career in physiology is the breeding ground for his thinking reflected in this book. As usual with his work, he dares to question general accepted theories and that alone makes this monograph worth reading. It will help the reader in developing his/her own independent scientific mind."

Jos AE Spaan
Professor Emeritus, Amsterdam University Medical Centres,
The Netherlands

"If, with René Descartes, one is convinced of the role of doubt and second-guessing in scientific-philosophical discourse, and that very unusual ideas should be discussed apart from the established mainstream, Noble has taken up the challenge with this book. In his new *work, he not only questions the standard concepts of electrophysiology and calls them inadequate, but also puts up for discussion a highly stimulating panorama of ideas about theories of physiological processes — including quantum physical considerations — in the whole of biology and challenges us to refute or validate them."*

Jochen Schaefer
Retired Professor,
International Institute for Theoretical Cardiology, Germany

"Scientific progress thrives on controversies. This book contributes significantly to that essential, ongoing process and discusses the nature of physiological mechanisms fundamental to life itself from an electron-based perspective. It is written by an established scientist in such a way that the reader is drawn into the debate. The content is accessible and of interest to a wide readership."

Gijs Elzinga
Retired Professor, The Netherlands

"Professor Noble takes a stab at explaining electrophysiology of excitable cells in terms of quantum physics. At the core of his daringly bold concept he replaces cation inflow as a cause for depolarisation of excitable cells with efflux of electrons provided by mitochondria. Even though the argumentation is yet to be supported by a similarly vast experimental body of evidence as conventional concepts of cellular electrophysiology, his book is a truly challenging, thought-provoking reading experience!"

Ursula Ravens
Senior Professor, Technische Universität Dresden, Germany

"Mark was able to explain complicated physiological phenomena in a comprehensive way. He is open-minded and likes new and challenging ideas."

Björn Wohlfart
Professor, Lund University, Sweden

Electromagnetism, Quanta,
and Electron Flow in the
Electrophysiology
of Living Cells

Other World Scientific Titles by the Author

The Cardiovascular System in Health and Disease
ISBN: 978-1-86094-278-5

Other Titles by the Author

The Cardiac Cycle
Cardiac Metabolism
Starling's Law of the Heart Revisited
The Interval-Force Relationship of the Heart: Bowditch Revisited
Mosby's Crash Course Cardiology (2 editions)
Snapshots of Hemodynamics (3 editions)

Electromagnetism, Quanta, and Electron Flow in the
Electrophysiology
of Living Cells

Mark I M Noble
University of Aberdeen, UK

World Scientific

EW JERSEY · LONDON · SINGAPORE · BEIJING · SHANGHAI · HONG KONG · TAIPEI · CHENNAI · TOKYO

Published by

World Scientific Publishing Co. Pte. Ltd.

5 Toh Tuck Link, Singapore 596224

USA office: 27 Warren Street, Suite 401-402, Hackensack, NJ 07601

UK office: 57 Shelton Street, Covent Garden, London WC2H 9HE

Library of Congress Cataloging-in-Publication Data

Names: Noble, Mark I. M., author.

Title: Electromagnetism, quanta, and electron flow in the electrophysiology of living cells / Mark I M Noble.

Description: New Jersey : World Scientific, [2021] | Includes bibliographical references and index.

Identifiers: LCCN 2021019302 | ISBN 9789811234941 (hardcover) | ISBN 9789811234958 (ebook for institutions) | ISBN 9789811234965 (ebook for individuals)

Subjects: MESH: Electrophysiological Phenomena | Electromagnetic Phenomena | Electrophysiology--methods

Classification: LCC QP341 | NLM QT 34 | DDC 612/.01427--dc23

LC record available at https://lccn.loc.gov/2021019302

British Library Cataloguing-in-Publication Data

A catalogue record for this book is available from the British Library.

For any available supplementary material, please visit
https://www.worldscientific.com/worldscibooks/10.1142/12229#t=suppl

Desk Editor: Shaun Tan Yi Jie

Typeset by Stallion Press
Email: enquiries@stallionpress.com

Acknowledgements

I am grateful to many scientific friends and colleagues for allowing me to argue with them about my theories. The Covid lockdown has curtailed these opportunities, so I am especially grateful to my wife, Prof Angela Drake-Holland, who is always available for scientific discussions!

About the Author

Dr. Mark I.M. Noble is Retired Honorary Professor of Cardiovascular Medicine at the University of Aberdeen, UK. He is an elected Fellow of the Royal College of Physicians of London, the European Society of Cardiology, the Royal Society of Medicine and the Association of Physicians. He has held the prestigious visiting professorships of Boerhaave Professor of Medicine at the University of Leiden and Spinola Professor of Medicine at the University of Amsterdam. His last position was Professor of Cardiovascular Medicine at National Heart and Lung Institute, Imperial College School of Medicine, London, UK. Since retirement and a move of his home to Scotland, he has acted as Honorary Professor of Cardiovascular Medicine at the University of Aberdeen, and has continued to publish high quality studies. He holds three doctorates (DSc, MD, PhD) from the University of London, UK.

Professor Noble is an international authority on a number of subjects within the field of the science of the cardiovascular system, with seven books and over 200 papers in international journals listed in PubMed covering a wide range of cardiovascular subjects. He is well known for definitive research results in cardiac muscle mechanics, global and regional heart functions and blood supply, the function of the coronary circulation, coronary arterial thrombosis, arterial haemodynamics, arterial endothelial function, arterial glycocalyx and its dysfunction in diabetes, prognostic risk factors on admission of patients with acute coronary syndromes, diagnostic and prognostic values of discharge of myocardial troponin T in acute coronary syndromes, the serotonin system in health and disease, platelet receptors and their antagonists, effects of cardiac denervation and transplantation, cardiac complication of AIDS, etc.

Preface

This book sprang from a request to me for the writing of a book on electrophysiology. The first question I asked myself was, "Is such a book necessary?" It was soon apparent that the answer was "yes", because the books available and the information on the internet seemed to me to be out of date in the interpretation of available data. The theories of science progress with the passing of human historic time. In physics, the universe was thought to be composed of:

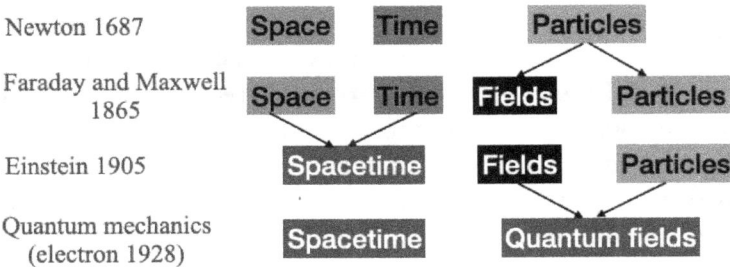

Newton 1687	Space	Time		Particles
Faraday and Maxwell 1865	Space	Time	Fields	Particles
Einstein 1905	Spacetime		Fields	Particles
Quantum mechanics (electron 1928)	Spacetime		Quantum fields	

I went to Grammar School in 1946 and Medical School in 1952. The physics that I was taught was Newtonian plus human use of electricity. But it is apparent that today physics is quantum mechanics. So my theme should be about quanta, of which the important ones in electrophysiology are the electrons.

But in quantum mechanics, it is not possible to say where an electron is, or what route it will take to go to another place. Feynman's integral is that the probability of going from A to B is the square modulus over all the paths of the exponential of classical action of the trajectory, multiplied by the imaginary unit and divided by Planck's constant!

This is not a language that I am competent to use to explain electro-physiology to anybody, but the present conventional theory of muscle, which is in terms of Newtonian, can at least be brought a bit up to the date of Faraday and Maxwell. The present conventional theory of cell electrical activation, which is in terms of electrophoresis, can be brought a little up to date by using the new 2019 definition of electricity.

I recommend physiologists to read the paper by Larrea Jiménez, "A quantum magnetic analogue to the critical point of water" (Jiménez, *2021*) in the April 2021 issue of *Nature*. Maybe they will share with me the realisation of how out of date our understanding and teaching are. I am a long way behind, but I hope this work will help people get into a different initial mindset.

Reference style

In the text I have not quoted more than two authors and the date of publication. The date is in italics so that if a reader is scanning the text, he/she can easily spot where there are references. The reference list is in alphabetical order of first author, which the reader will know from the text. I prefer this system to numbered references. With that system one comes across a number in the text, then one has to go to the reference list to find out who is being quoted. I hope readers will find my system less disruptive to the flow of the narrative.

Contents

Acknowledgements   vii

About the Author   ix

Preface   xi

Abbreviations   xvii

[A] Introduction   1

Chapter 1. Facts and definitions   5

Chapter 2. What about volts?   7

Chapter 3. What is the current idea about cellular voltage?   10

Chapter 4. The concept of electron density   14

Chapter 5. Does the electrolyte distribution actually
            determine the trans-membrane potential?   16

Chapter 6. Where, in living cells, is electricity generated?   18

Chapter 7. Depolarisation (loss of cytoplasmic electron
            density) activation   22

Chapter 8. A test of the impedance hypothesis   26

Chapter 9. Fast depolarisation in conducting tissues   28

Chapter 10. Repolarisation   31

Chapter 11. The calcium, sodium and potassium problems   35

Chapter 12. Excitation-contraction coupling in muscles   43

Chapter 13. The effect of the Ca^{2+} problem on cardiac
            muscle electrophysiology   46

Chapter 14. Mechanical restitution and the optimal
            contractile response   49

Chapter 15. Internal calcium ion release and recirculation   53

[B] *Non-Electromagnetic Theory versus
     Electromagnetic Theory of Muscular Function*

Chapter 16. Objections to the non-electromagnetic
            theory of striated muscle   64

Chapter 17. Different theories   71

Chapter 18. Electromagnetic theory of muscle contraction   73

Chapter 19. Electrophysiology of smooth muscle   86

Chapter 20. Flow-mediated dilatation   90

Chapter 21. Pulmonary vessels   95

Chapter 22. Non-vascular smooth muscle   97

Chapter 23. Hierarchy of vertebrate muscle   104

[C] *Other Organs*

Chapter 24. Electrophysiology of endocrine glands   106

Chapter 25. Electrophysiology of exocrine glands   110

Chapter 26. Exceptions to any general model   112

Chapter 27. Central and autonomic nervous systems   115

Chapter 28. Receptors affecting perception   119

Chapter 29. Receptors initiating feedback control reflexes   128

Chapter 30. Summary and general comments
            on vertebrate animals   136

[D] *Invertebrates*

Chapter 31. The electrophysiology of invertebrates   138

[E] *Interlude*

Chapter 32. Electricity passing through flesh   151

[F] *Other Organisms*

Chapter 33. Plants   158

Chapter 34. Fungi   167

Chapter 35. "Primitive" organisms   168

Chapter 36. Bacteria   173

Epilogue   181

References   185

Index   197

Abbreviations

A	ampere
AC	alternating current
ACh	acetylcholine
ACTH	adrenocorticotropic hormone
ATP	adenosine triphosphate
ATPase	enzyme that splits ATP to ADP (adenosine diphosphate) and P (phosphate)
Ca^{2+}	calcium ion
Cyclic AMP	cyclic adenosine monophosphate
DC	direct current
DNA	deoxyribonucleic acid
ECG	electrocardiograph
EDRF	endothelium-dependent relaxing factor = nitric oxide
EMG	electromyograph
ENG	electroneurograph
G-protein	guanine nucleotide-binding proteins are a family of proteins that act as molecular switches inside cells, and are involved in transmitting signals from a variety of stimuli outside a cell to its interior.
GABA	gamma-aminobutyric acid
H^+	hydrogen ion
$[H^+]$	hydrogen ion activity
IP3	inositol triphosphate or inositol

K^+	potassium ion
$[K_i^+]$	intracellular potassium ion concentration
$[K_o^+]$	extracellular potassium ion concentration
kg	kilogram
ln	logarithm to the base e (natural logarithm)
log	logarithm to the base 10
min	minute
mg	milligram
mM	millimolar
Mn^{2+}	manganese ion
MRI	magnetic resonance imaging
msec or ms	millisecond
mV	millivolt
Na^+	sodium ion
Na^+/K^+ ATPase	sodium pump (expels 3 Na^+ out of cell for 2 K^+ in)
NCX	Na^+/Ca^{2+} exchanger (expels 1 Ca^{2+} out of cell for 3 Na^+ in)
pH	$-\log [H^+]$
QT interval	time between the Q wave and the T wave of the ECG
s	second
SI	Systeme Internationale
SL	sarcomere length
SR	sarcoplasmic reticulum
VSM	vascular smooth muscle

Introduction

Electrons, which are sub-atomic particles that can also act as electromagnetic waves, are ubiquitous in the universe. All atoms have an outer layer of electrons orbiting the nucleus. Thus all matter, consisting of atoms or ions, contain electrons. On planet Earth, the existence of lightning shows that electrons can also flow freely outside atoms. Thomas Edison and Nikola Tesla were prominent in the early understanding of electricity in the nineteenth century. Electron flow is usually called electricity, but electricity is only part of electromagnetism. Humans discovered that they could generate electricity by moving magnets and use this generated electricity for useful purposes by energy conversion. For instance, Michael Faraday invented the electric motor in 1821, and Georg Ohm mathematically analysed the electrical circuit in 1827. Electricity, magnetism and light were definitively linked by James Clerk Maxwell (Figure 1) in 1862.

It is claimed, "Physiologists have taken up the quantum baton, becoming increasingly interested in the fusion of 'quantum physics' with the 'physiology' of life molecules to provide answers where traditional classical approaches have consistently failed (Wolynes, 2009; Natochin, 2010)." That does not seem to be the case in electrophysiology, where the role of electrons (quanta) has been neglected, even condemned.

Most of our experiences of electricity (electrons moving) arise from human applications, which include its use in medicine. The early pioneers thought there were two kinds of electricity — arbitrarily named

Figure 1:   James Clerk Maxwell. Source: Wikimedia Commons, public domain.

positive and negative. We now know that there is only "negative" electricity, which is logically absurd, but the consequence is that the electron, the fundamental particle of electricity, still, by convention, has a negative sign. This certainly does not matter when considering the distribution of electricity by humans because alternating current (AC) is used for ease of transforming voltages; the electrons are going forwards and backwards usually 50 times a second. There is no AC in natural phenomena. It is different with direct current (DC); e.g., switch a torch battery on, and electrons pass from the "negative" terminal to the "positive" terminal. Yet it still does not matter if the user erroneously thinks that current is going from the "positive" terminal to the "negative" terminal, because the torch light comes on. The same is true for a whole range of DC-powered devices.

When it comes to the electricity in life, this confusion may matter. That electricity is a characteristic of life forms is obvious when one considers the shock that an electric eel can generate to kill its prey, and

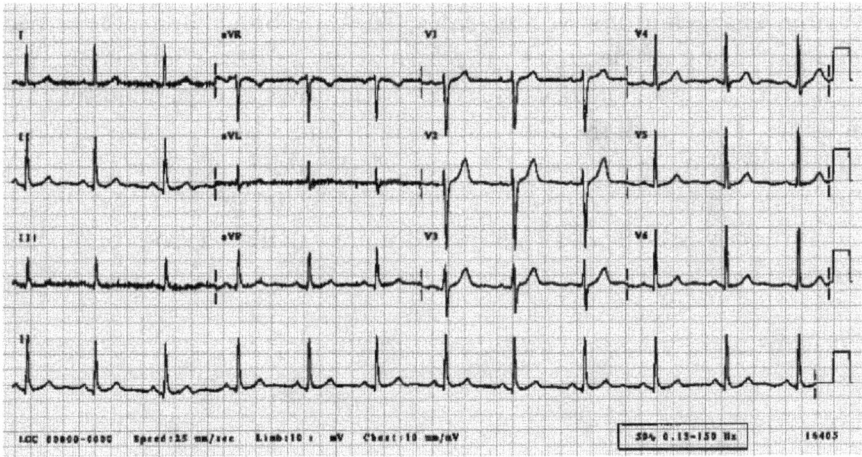

Figure 2:   Human ECG (the author's).

the electric impulses generated by the roots of mushrooms growing on wood. The whole story is huge.

The measurements with electrodes and amplifiers of the electric voltages on the surface of the human body in diagnosis include the electrocardiograph (ECG, illustrated in Figure 2) in diagnosis of heart diseases, electroencephalograph (EEG) in diagnosis of brain diseases, electroneurograph (ENG) in diagnosis of peripheral nerve diseases and electromyography (EMG) in diagnosis of muscle diseases.

It turns out that the wrong polarity of conventional labelling of electricity in these applications does not matter; the correct clinical conclusions still apply. The same is true for the use of electricity in medical treatments. During my career, mostly in cardiology, there were great advances in these treatments, e.g., pacemaking, defibrillation, and ablation in arrhythmia therapy. These are all effective whether the cardiologist thinks in terms of electrons with a negative polarity or something else with a positive polarity.

Where polarity does matter is when we try to understand the electrical phenomena in living organisms and living cells. Living organisms are called organic because they are constructed mainly of organic molecules, identified by chains of carbon atoms with attachments of various charged atoms or molecules. These carbon chains are formed by the sharing of electrons. The formation and function of organic molecules are possible because of electron flows as explained by Paul Scudder. The title of the present monograph mirrors his classical work "Electron Flow in Organic Chemistry" (Scudder, 2013) and applies the same principles to electron flows within living cells. My electrical authority is Bird (2014). Some data are calculated from published works.

CHAPTER 1

Facts and definitions

Fact: Electrical voltages are recorded on the surface of organisms, e.g., electrocardiograph (ECG) in the human, i.e., the organism generates electricity.

Definition: The Systeme Internationale (SI) re-defined all base units in 2019. In the case of electricity, **redefined as electrons moving**, the base unit is the ampere, defined as: "The ampere, symbol A, is the SI unit of electric current. It is defined by taking the fixed numerical value of elementary charge e (electron) to be $1.602176634 \times 10^{-19}$ when expressed in the unit C, which is equal to A·s, where the second (s) is defined in terms of $\Delta \nu_{Cs}$."

Fact: The mass of an electron is 9.109×10^{-31} kg. Electrons are extremely small. The mass of an electron is only about 1/2000 the mass of a proton or neutron, so electrons contribute virtually nothing to the total mass of an atom. As a sub-atomic particle, they follow quantum mechanical rules, not Newtonian rules. The great scientist Richard Feynman was quoted as saying, "I think I can safely say that nobody understands quantum mechanics" (Feynman, *2011*); maybe the reader should try Rovelli (*2014*), who tries to explain it. The Bohr model of the atom, in which electrons orbit about the nucleus, is a convenient representation, but it is not complete; electrons within the atoms act more like electromagnetic waves than orbiting particles. According to a basic principle of quantum mechanics, it is impossible to know simultaneously both the exact position and momentum of an electron; this means that the trajectory of the electron cannot be precisely determined. Nevertheless, considering free electrons as particles is adequate for many applications.

Fact: The working current in a circuit, e.g., battery-operated torch, conventionally assumed to be a passage of electricity from the positive to the negative pole, is actually a flow of electrons from the negative to the positive pole.

Implication: A supposed inward current into a living cell may actually be an outflow of electrons. A supposed outward current from a cell may actually be an inflow of electrons. There are also ion flows of which calcium ion (Ca^{2+}) current is of great importance, as well as diffusion of charged molecules.

There are other important mechanisms that appear to be essential accompaniments of this electrical phenomenon, particularly the calcium inward current, the sodium/calcium exchanger and the Na^+/K^+ ATPase (sodium pump). I will argue that there must be a mechanism for K^+ extrusion, but this is not clear to me.

The purpose of the present work is to re-interpret electrophysiology in terms of these implications.

CHAPTER 2

What about volts?

The reader will be familiar with voltage, the electric potential difference between two points, e.g., between the poles of a battery. 1 volt = 1 joule (of work) per 1 coulomb of electron charge. The voltage between the inside and the outside of living cells is negative using the conventional arbitrary polarity, e.g., −80 millivolts (mV) for a nerve (this is called the trans-membrane potential). As the electron, according to conventional arbitrary polarity, has been assigned as negative, a negative voltage of a living cell implies an excess of free electrons over and above those circulating in orbit around the nuclei of the atoms and ions. An alternative explanation would be if there were more negatively charged than positively charged molecules and ions in cells. This was disproved by Anderson (2013) who found that the total positive and negative charges inside cells were equal, i.e., there is no excess of negatively charged molecules and ions (Figure 3).

Implication: **The conventional intracellular negative voltage must be due to electron excess.**

So what determines the negative voltage of cells? This is when authors referred to the Nernst equation, but did not quote the original Nernst equation (1), which only describes a *change* of trans-membrane poten-

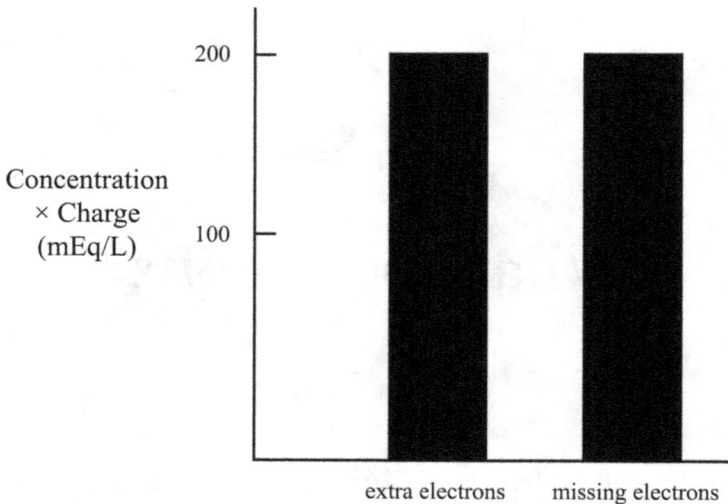

Figure 3: The left column represents all the so-called "negative" intracellular charges (anions: atoms and molecules with extra electrons in the valence orbital). The right column represents the so-called "positive" intracellular charges (cations: atoms and molecules with missing electrons in the valence orbital). Not previously published.

tial, not the actual trans-membrane potential:

$$E_{cell} = E_0 - (RT/nF)lnQ \tag{1}$$

where:

E_{cell} = cell potential under non-standard conditions;

E_0 = cell potential under standard conditions, in which there is a negative trans-membrane potential, i.e., a net excess of intracellular negative electrical charges;

R = gas constant which is 8.31 (volt·coulomb)/(mol·K);

T = temperature in Kelvins;

F = Faraday's constant, 96,500 coulombs/mol;

Q = Reaction quotient, which is the equilibrium expression with initial concentrations rather than equilibrium concentrations;

n = number of moles of electrons exchanged in the change from E_0 to E_{cell}, i.e., an equation to evaluate the difference in electric potential between E_{cell} and E_0.

If we consider an organism with a constant temperature, e.g., a human with no fever or hypothermia, R, T, F and Q are constants, i.e., parameters of the equation. Therefore, in Equation (1), E_{cell} is the dependent variable and n is the independent variable. In the case of fever, hypothermia and hibernation, T becomes a second independent variable affecting cell potential.

Implication: **The trans-membrane potential at any time depends on the number of electrons lost or gained from the resting control value.**

CHAPTER 3

What is the current idea about cellular voltage?

The version of the Nernst equation quoted in most texts is quite different from the original (Equation (1)).

A Nernst equation presented as one for the absolute value of E_m is:

$$E_m = \frac{RT}{F} \ln \left(\frac{r P_{K^+}[K^+]_{out} + P_{Na^+}[Na^+]_{out}}{r P_{K^+}[K^+]_{in} + P_{Na^+}[Na^+]_{in}} \right) \qquad (2)$$

where:

E_m = the membrane potential (in volts, equivalent to joules per coulomb);

P_{ion} = the permeability for that ion (in metres per second);

$[ion]_{out}$ = the extracellular concentration of that ion;

$[ion]_{in}$ = the intracellular concentration of that ion in the same units as $[ion]_{out}$ (absolute units do not matter, as the ion concentration terms become a dimensionless ratio);

R = the ideal gas constant (joules per Kelvin per mole);

T = the temperature in Kelvins;

F = Faraday's constant (coulombs per mole);

r = the absolute value of the transport ratio (1.5 in the case of the Na^+/K^+ ATPase sodium pump).

With this equation, the resting stable trans-membrane potential depends on the distribution of electrolytes: potassium ion (K^+) and sodium ion (Na^+) concentrations outside the cell in the extracellular fluid, and inside the cell. This ion distribution is supposed to be maintained by the sodium pump. The cell membrane which covers the surface of a cell consists of two sheets of lipid cells, called the inner and outer layers of a lipid bilayer. This separates the inner medium of the cell from the extracellular fluid. The material inside the membrane was squeezed out and found to be like a jelly (a gel) and was called **protoplasm**. In spite of this, later investigators insisted that the intracellular compartment was a liquid solution that they called **cytosol**. However, modern research using magnetic resonance has established that:

Fact: **The intracellular material is indeed a gel in which water (the main molecule) is structured.**

This jelly-like material is now called the **cytoplasm**. 60% of the human body and 94% of cucumber are water, but both have structure due to the intracellular water being structured in a gel. Gels conduct electricity, i.e., electrons flow through gels. The main objection to my theory of electron flow, from the opinion leaders in electrophysiology, is, "Our definitive criticism is that significant free-electron movement in an aqueous ionic medium cannot occur." And the electric eel (Figure 4) lives in an aqueous ionic medium!

Figure 4:   An electric eel (FMIB 38638 *Gymnotus electricus*). Source: Wikimedia Commons, public domain.

All fish can sense electricity and some species can communicate with each other using electricity. And they live in an aqueous ionic medium! I conclude that the statement, "Our definitive criticism is that significant free-electron movement in an aqueous ionic medium cannot occur", is wrong. In any case, why attack a phenomenon that I never postulated anyway? I postulated electron flow through **gels**. I am surprised that anyone still thinks the intracellular compartment is liquid.

Side issue: Humans, when they mastered commercial electricity with its many uses, created the process of electrolysis. Two electrodes were inserted into a bath of electrolyte solution, e.g., sodium chloride (salt). One, called an anode, was connected to a positive voltage and the other, called a cathode, to a negative voltage so that a current passed through the solution. It was found that the negatively charged ions accumulated on the anode and the positively charged ions on the cathode. (The process was adapted to create metal plating.) This led to negatively charged ions being called anions and positively charged ions to be called cations. There is no electrolysis in nature, but the nomenclature was adopted in biology and affected the nomenclature of large organic molecules, so that we have anionic proteins and anionic lipids. This is important when reconsidering Anderson's data. Although the total positive charges equalled the negative charges, the negatively charged molecules were large, mostly anionic proteins, whereas the positively charged molecules were mostly potassium ions (K^+). As every cook knows, one prepares a jelly by dissolving protein in water. The jelly-like nature of the intracellular compartment is explained by the high protein content, and as these are mostly negatively charged, the K^+ are bound to them electrostatically. With reference to "What is the current idea about cellular voltage?", because of the assumptions in Equation (2), it is dependent on the whole Na^+ and K^+ distribution between the extracellular and intracellular compartments being maintained continuously by the sodium pump.

The sodium pump is an enzyme protein in the cell membrane that extrudes 3 Na^+ to the extracellular fluid in exchange for taking 2 K^+ into the cell.

This explains the transport ratio of 1.5 in equation 2 above (3 divided by 2 equals 1.5). The sodium pump is more properly termed the Na^+/K^+ ATPase, because it consumes energy by de-phosphorylation of adenosine triphosphate (ATP) produced mainly in mitochondria via the process of oxidative phosphorylation. (Some non-oxidative phosphorylation to produce ATP can also occur during the breakdown of glucose.)

CHAPTER 4

The concept of electron density

The difficult aspect, in which the wrong assignation of a negative sign to electrons occurs, is when considering cell electrophysiology. For example, taking the cell with the most negative cytoplasmic transmembrane potential, the heart ventricular cell: with an external potential of zero and cytoplasmic potential of $-85\,\text{mV}$, it is natural, if one assumes the conventional polarity, to expect a current going from the higher voltage of zero to the lower voltage of $-85\,\text{mV}$. Indeed, that is what is assumed in most texts, and as such "inward" current is sodium ion (Na^+) dependent, the inward current was assumed to be carried by Na^+. Scudder (2013) had a similar problem in analysing electron flow in organic chemistry, and, as a result, turned to the concept of electron density.

Definition: Electron density is the *probability* that electrons occur in a certain location.

Electrons are never static. They are always on the move, in and out of atoms and ions, and can be free, causing, in the case of cellular cytoplasm, a high electron density (conventionally a negative electric potential). Electron density therefore has a positive stochastic nature. Scudder (2013) went on to state the principle that electrons flow from locations of higher electron density to locations of lower electron density. Now, going back to our example of the heart ventricular muscle, it is natural to think that electrons will flow out of the high electron density cytoplasm into the low electron density extracellular space. In order to use an index of electron density, I use the root mean square of the electric potential, e.g., -8 mV squared is $+64$ mV^2; the square root of $+64$ mV^2 is $+8$ mV. This is convenient because the index is simply the conventional electric potential with the sign positive instead of negative.

In conventional DC electricity, Ohm's Law gives current as voltage over resistance. More correctly, current equals voltage/impedance. Impedance encompasses all factors impeding the flow of electrons including resistive, capacitative and inductive impedance, and semi-conduction. Now, we can say that in living cells, electric current is determined by the difference in electron density (between the locations being considered) divided by the impedance of the intervening matter. Thus, if the impedance drops or the difference in electron density increases, more electrons will flow, and vice versa. The potential importance of this relationship will arise in a later section.

CHAPTER 5

Does the electrolyte distribution actually determine the trans-membrane potential?

With the adoption of the electron density concept, we are rid of the problem in plotting negative numbers. The main independent variable in the much used Equation (2) is the ratio of the external potassium ion concentration ($[K_o^+]$) to the intracellular potassium ion concentration ($[K_i^+]$).

In Figure 5, the electron density of the cells of various tissues is plotted against this ratio. The tissues were heart ventricle (highest electron density of 85 mV), skeletal muscle, vascular endothelium, nerve, adrenal cortex, leukocyte, salivary gland acinar cell, sinus node cell, vascular smooth muscle cell, platelet, brown fat cell, retinal cell, pancreatic acinar cell, liver, fat cell, pancreatic islet cell, and red blood cell (lowest electron density of 9 mV). All of these tissue cells have the same $[K_o^+]/K_i^+]$, according to the table of Milo & Philips (2020), so it is not necessary to perform statistical correlation analysis to conclude that Equation (2) is wrong. Trans-membrane potential and electron density are not determined by $[K_o^+]/K_i^+]$.

There are small variations in $[K_o^+]/K_i^+]$ of different tissues but they do not affect this conclusion. For example, taking the cells with the

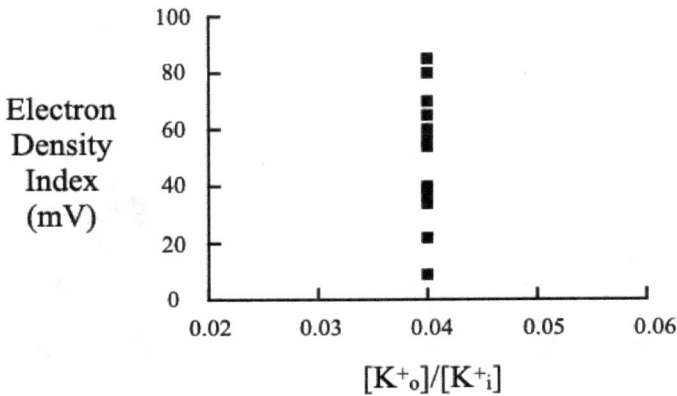

Figure 5: The electron density and trans-membrane potential are not related to the ratio of potassium ion concentration outside the cell to that inside. Each solid square comes from a different tissue within the same organism. Not previously published.

highest and lowest electron density, the heart ventricle has an electron density of about 85 mV, while the red blood cell has an electron density of only 9 mV. The mature red blood cell (the simplest cell in the body) has lost its sodium pump during maturation. The highest estimate of red blood cell sodium concentration is 11.4 ± 3.1 mM. This must be compared with the extracellular sodium concentration of 145 mM. The intracellular potassium concentration of the red blood cell is 80–120 mM compared to 3.5–5 mM concentration in plasma. So the red blood cell can achieve a normal low intracellular sodium concentration and high intracellular potassium concentration without a sodium pump (or a very weak one in immature red blood cells).

Conclusion: **sodium pumping by the Na^+/K^+ ATPase is not the cause of the low intracellular Na^+ and high intracellular K^+ that is assumed to determine trans-membrane potential, and therefore electron density.**

As will be discussed in a later section, the function of the sodium pump is to extrude Na^+ when it increases *above* the control concentration.

CHAPTER 6

Where, in living cells, is electricity generated?

The cells of the body differ widely from organ to organ and tissue to tissue but Figure 6 shows a blueprint for the composition of cells. The nucleus contains the genes (DNA) which determine what's what, including possibly the intracellular ionic concentrations. For my story I would like the reader to focus first on the cytoplasm. It is important to emphasise again that the cytoplasm is a gel.

When an electrode is placed in a cell, one measures the electron density of the cytoplasm, e.g., 70 millivolts (-70 mV electric potential difference in the old money). I also want to focus on the cell membrane which enfolds the rest of the cell. This is a double layer of lipid molecules, each layer one molecule thick. The outer layer is electrically neutral but within the inner layer there are lipids charged with extra electrons (so-called anionic phospholipids), electrons that are shared with calcium ions (Ca^{2+}). These ions cling to the inner layer by a mechanism called electrostatic binding. Then I want you to see that there are other things within the cell; these are called organelles, e.g., nucleus and mitochondria.

In Figure 7, the nuclei are stained purple and the mitochondria yellow.

Then I searched the literature to find the voltages of some of the organelles and converted them to electron density values. I thought that the most likely of these organelles to be the electricity generator

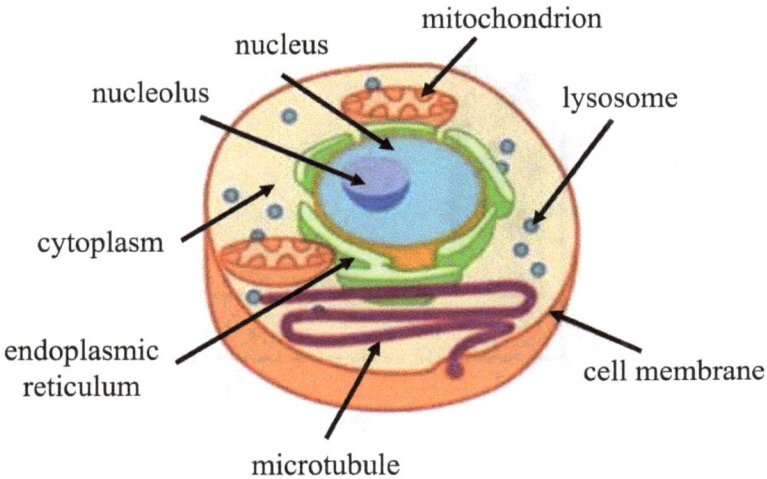

Figure 6: Schematic diagram to illustrate the basic constituents of a generalised cell. Source: Adapted from drawing MeSH D002477, TH H1.00.01.0.00001, FMA 686465, Animal terminology [edit on Licenced Wikidata], and labelled by me.

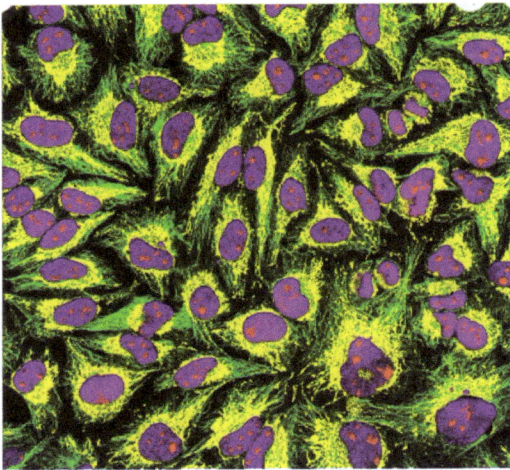

Figure 7: HeLa cells grown in cell culture and processed to reveal microtubules with an antibody to β-Tubulin (green), mitochondria with an antibody to Heat Shock Protein 60 (HSP60, in yellow), nucleoli with an antibody to Fibrillarin (red) and nuclear DNA (purple). Cells grown and processed and all antibodies used were generated by scientists at EnCor Biotechnology Inc. Licenced under Creative Commons Attribution-Share Alike 4.0.

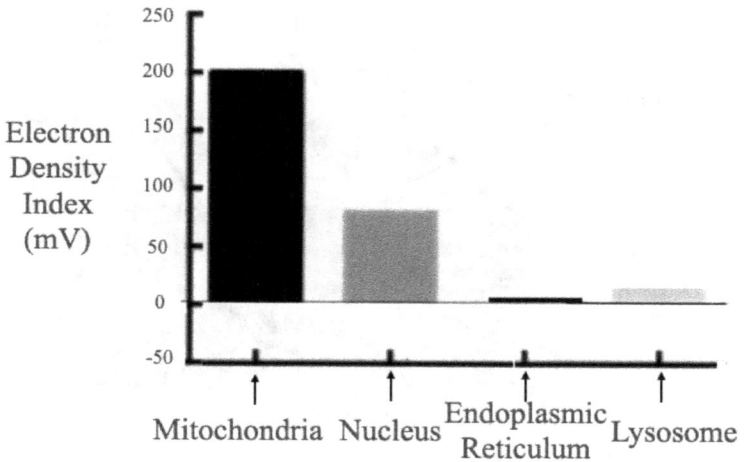

Figure 8: Electron densities of four intracellular organelles. Not previously published.

would be the one with the highest electron density. I have not yet obtained data on some of them, but four are illustrated in Figure 8.

I need to look for the electron density of the other organelles in case there is one with an electron density greater than that of mitochondria, but for the time being, the mitochondria have by far the greatest electron density of those looked for. Therefore, I suggest that the mitochondria are the electricity generators of living cells. Every cell in our bodies contains these microscopic biological machines which convert molecules broken down from food to **chemical energy** in the form of adenosine triphosphate (ATP), and **electrical energy** (electrons).

There was once a question on BBC's University Challenge programme, "What is the biochemical process that separates protons from electrons?" Normally protons, which are hydrogen ions (hydrogen atoms missing an electron), combine with electrons to form hydrogen atoms. Answer: "oxidative phosphorylation". Oxidative phosphorylation takes place in mitochondria. How do the mitochondria get rid of the separated protons? By means of an active pump that generates a proton concentration gradient across the inner mitochondrial membrane, because there are more protons outside the matrix than inside. In the intracellular compartment the protons from the mitochondria add to those in the cytoplasm which determine the degree of acidity or alkalinity of the medium. (pH = the negative log-arithm to base 10 of the hydrogen ion concentration, or $-\log [H^+]$.) As protons carry an electric charge, one can anticipate that pH will have an effect on electrophysiology; this will be illustrated in a later section.

Further, I suggest that, as with organic molecules, electrons flow from the high electron density sites (mitochondria) to other parts of the cell with lower electron density (e.g., the cytoplasm) and from cytoplasm to the extracellular compartment.

CHAPTER 7

Depolarisation (loss of cytoplasmic electron density) activation

The best example of spontaneous depolarisation is found perhaps between the action potentials (the diastolic interval — double arrow in Figure 9) of the sinus node cell. The sinus node of the heart is the clock that initiates the heartbeat for a lifetime. It is very important because life depends on the mammalian heart beating.

Illustrated in Figure 9 are the electrical events in a sinus node cell. The diastolic interval, indicated by the double arrow in the picture, shows a drift. The upper trace shows how it is depicted in publications using the conventional polarity. It illustrates an upward drift of trans-membrane potential. As a result of the widespread opinion that electricity consists of something positively charged going to a less positively charged location, this has been interpreted as being caused by an inward current carried by a positively charged ion (cation). Actually no such ion flow has been identified. However, the persistence of the idea of electric current being positive results in this current now being known as the "funny" current (I_f, I indicating current and f indicating funny). This concept is elaborated by the thought that there must be an unknown ion entering the cell through an ion channel.

By contrast, the lower trace shows the same phenomenon in terms of change in electron density. Electron density slowly decreases during

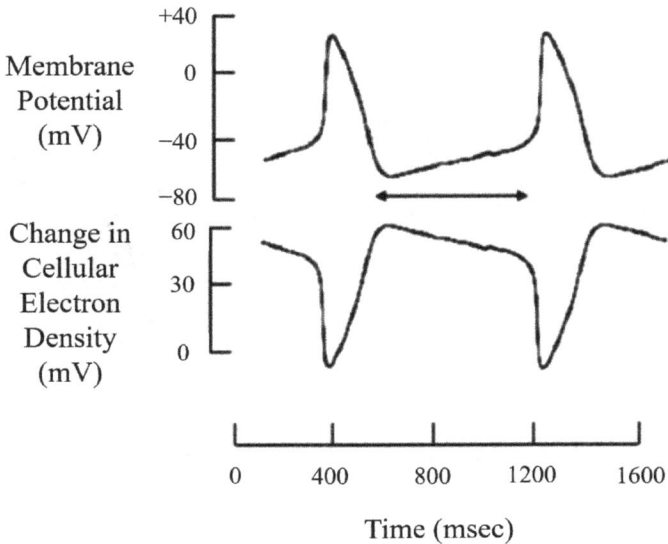

Figure 9:   The top trace is an original trace of trans-membrane potential of a sinus node cell. The lower trace is the calculated electron density. The double arrow indicates diastole. Not previously published.

the diastolic interval. This is straightforward in interpretation as an electron flow from the cytoplasm to the extracellular space. There is no need to invent an imaginary ion. The reason for the ordinary DC outward current can be postulated to be that the intervening impedance is insufficiently great to prevent some electron flow from the 60 mV electron-dense intracellular source to an electron density of 40 mV.

A relevant experiment on sinus node cells by Moe's group (Jalife, *1980*) is worth re-interpreting in the light of the change in polarity from positive electricity to electron flow. I achieved this by inverting the published traces, as in the lower panel of Figure 9, and reversing the meaning of the pulses of electric current that were injected, as pulses of added electrons.

In the upper panel of Figure 10, the re-interpreted tracings are presented as electron density index, so that at the beginning of the diastolic interval, the trans-membrane potential of −60 mV becomes an electron density index of 60 mV. With increasing amplitudes of injected electrons 1 to 4, the diastolic interval becomes progressively shorter in duration. By adding electrons the electron density during

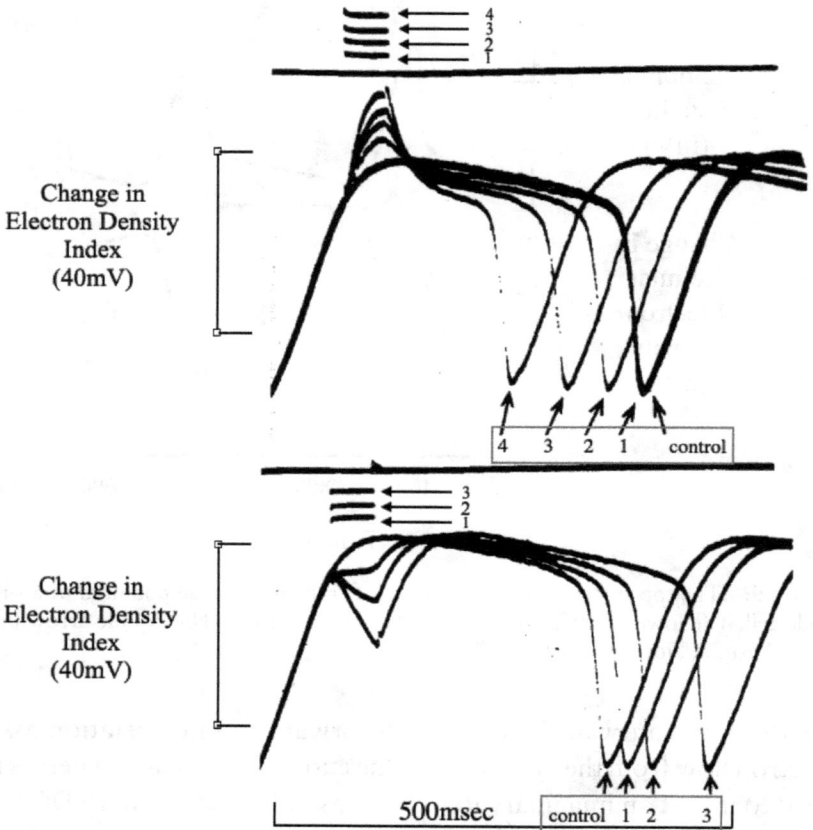

Figure 10: Effect of injecting and withdrawing electrons in the sinus node cell. Numbers in the upper panel indicate increasing numbers of electrons injected and in the lower panel the increasing numbers of electrons withdrawn. Not previously published.

the diastolic interval reaches the threshold at 40 mV earlier for the firing of the next action potential. When electrons are withdrawn from the cell (1–3 lower panel), the diastolic interval becomes progressively longer in duration as the outflow of electrons is retarded. This re-interpretation of the experiment does not prove my electron theory because theories cannot be proved, only disproved, but doing the experiment by injecting and withdrawing funny current is not possible as nobody knows what is the positively charged ion of which it is composed.

It has been noted that when the electron density has drifted down to approximately 40 mV in the sinus node cell, there is a sudden, more rapid fall in electron density which triggers electrical excitation of the heart, followed by contraction. However, at this point, I want to draw attention to another principle that I have followed. An illustration of this is that sometimes there is a need for the heart rate to increase in frequency and this is achieved by shortening of the duration of the diastolic interval (double arrow in Figure 9). This is realized in turn by an increase in the rate of diastolic electron flow, mediated by a lowering of the diastolic impedance. This is then accomplished chemically by either release of noradrenaline from sympathetic nerve endings, or by an increase in the concentration of adrenaline in the blood (secreted from the adrenal gland). Both these chemicals exert their effect by acting upon receptors on the cell membrane. It is generally thought that noradrenaline and adrenaline bind to adrenoreceptors on the surface of the cell. The activated receptor binds a G protein on the inside of the membrane. The activated G protein causes the enzyme adenylyl cyclase to convert ATP to cyclic adenosine monophosphate (cAMP), called the second messenger. I suggest that the end result is a decrease in electrical impedance. Another example, still with the sino-atrial cell, is the slowing of heart rate by increase of diastolic interval via acetylcholine released from parasympathetic vagal nerve endings.

Principle: **Depolarisation resulting from interaction of chemicals (neurotransmitters, hormones and cytokines) with receptors on the cell surface followed by a chain of reactions involving second messengers have the end result of decreasing electrical impedance, thus allowing increased electron outflow. Much less frequently, hyperpolarisation (i.e., increased electron density) results from interaction of chemicals with receptors on the cell surface followed by a chain of reactions involving second messengers having the end result of increasing electrical impedance, thus producing decreased electron outflow.**

In the latter case electrons from the mitochondria (see later section on repolarisation) accumulate in the cytoplasm.

| CHAPTER 8 |

A test of the impedance hypothesis

This was tested by Silvio Weidmann (*1951*) in the Purkinje fibre. These fibres consist of specialised cardiac muscles that lie on the inner surface of the heart ventricles and conduct electricity to the main contracting muscle. They have a diastolic electron outflow which is much slower than that of the sinus node cell, but can initiate a slow ventricular beat if the sinus node is out of action in heart diseases. Figure 11 is derived from his paper in which he measured changes in cytoplasmic trans-membrane potential. I have changed the white-on-black photograph of his cathode ray oscilloscope to a black-on-white picture and reversed trans-membrane potential trace to obtain the index of electron density. The spike was Silvio's term for the upstroke of the action potential, which was so fast that even his oscilloscope sensitivity could not show it; I have drawn a line where it should be. This is a very fast loss of electrons, of which, more later.

In order to obtain an index of impedance, he imposed constant amplitude pulses of electric current, causing varying vertical deflections on top of the natural change in electron density of the action potential, i.e., the vertical thickness of the trace is an index of impedance.

The thickness during the spike is so thin that it cannot be interpreted as a detectable difference from zero. Hence, the loss of electrons is maximal as indicated by the inserted straight line.

The Spike

Change in
Electron Density
(50mV)

400 msec

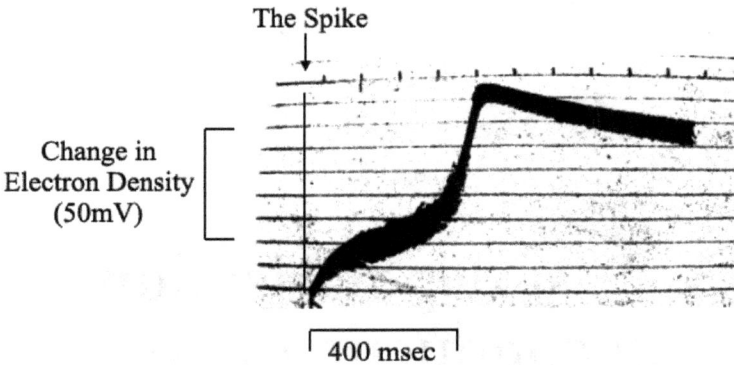

Figure 11:   Original construction by the author to illustrate electron density changes. A Purkinje fibre action potential with imposed constant amplitude current pulses so that the vertical thickness of the electron density trace is an index of electric impedance. Not previously published.

The thickness of the trace is maximal during what is called the plateau of the action potential, although, in fact, the overall electron density is increasing. The inference from the maximal thickness between 200 and 300 milliseconds (msec) after the spike is that electron outflow is minimal, if indeed it is not zero. After the plateau, there is an increased rate of increase of electron density to the highest value at the beginning of diastole. This is called repolarisation and I will, later in the work, postulate that electrons flow from the mitochondria into the cytoplasm at this time. Note that during the following diastole, the vertical width of the trace is less than that during the plateau, indicating a lesser impedance, which allows an overall downward drift of the trace. This is the outflow of electrons of the pacemaker current.

Conclusion: **Such evidence as is available is compatible with the hypothesis that electron flow is subject to changes in electrical impedance.**

| CHAPTER 9 |

Fast depolarisation
in conducting tissues

Just before Weidmann's experiment, re-analysed in Figure 11, he and his colleague Draper (*1951*) had recorded the action potential of the main working ventricular cells. In Figure 12, I have treated their trace in the same way as Figure 11. Once again I have added a line to indicate the spike, which could not easily be discerned on the published oscilloscope photo reproduction.

Why does a spike of very rapid depolarisation occur? Could the cell be short-circuited by the immediate proximal cell depolarisation? As most householders know, a short circuit causes such a large flow of electrons that it breaks a fuse. It is not possible from this recording to determine how large the electron flow is during the spike, but this has been measured by the group in Amsterdam, and found to be 430 volts/sec (Berecki, *2010*).

The action potential is conducted, spreading throughout the ventricular myocardium, acting as a syncitium. I had an idea that a simple explanation is to interpret this as characteristic of a propagating electrical short circuit. A more complex method of conduction, that is widespread in tissues, is via gap junctions. Gap junctions are clusters ("plaques") of hydrophilic intercellular channels formed by the docking of two ion channels or connexions, each contributed by a neighbouring cell, allowing the direct transfer of signalling molecules and providing a pathway of low resistance for the spread of electrical

The Spike

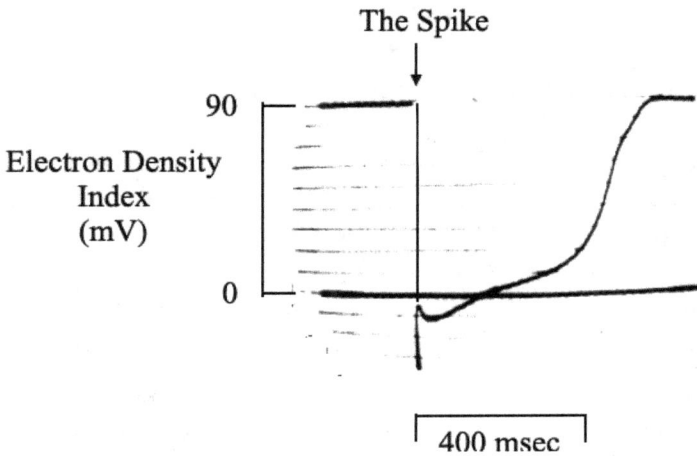

Figure 12:   Cardiac ventricular action potential subject to calculation by the author to show electron density changes. Not previously published.

currents between cells. It is thought that gap junctions are particularly important in cardiac muscle: the signal to contract is passed efficiently through gap junctions, allowing the heart muscle cells to contract in unison. Whether this is a fast enough system to account for the rapid spread of depolarisation in ventricular muscle remains a question, but it is a widespread mechanism in many tissues.

The mechanism of conduction spread through the cardiac muscle does not answer the question, "Why does a spike of very rapid depolarisation occur?" As one can observe from the title of the article of Berecki and colleagues (2010), even in 2010, they called this the velocity of sodium ion (Na^+) inward current rather than the velocity of electron outflow. In my opinion, an Na^+ inward current is not possible because the sodium ion is 10,000 times the mass of an electron. The attribution to a Na^+ current derives from Hodgkin and Huxley's (1952) original finding in isolated squid giant axons.

These were great physiologists that I knew and admired as a PhD student in the early 1960s. They amputated the giant axon of the squid, removed the cytoplasm and organelles and replaced them with a physiological solution into which an electrode was placed. Electric current was passed between this and an external electrode. Inward current (conventional polarity, but actually electron outflow) was dependent on the presence of Na^+ in the external solution, while outward current (actually electrons forced into the axon, which never happens in nature to my knowledge) was dependent on the presence of potassium ions (K^+) in the artificial intracellular fluid. These were brilliantly performed experiments in the 1950s and were very influential. It seems that physiologists have extrapolated these findings to an assumption that Na^+ carries the depolarisation of the spike and K^+ carries the repolarising current, and applied it to all conducting tissues including heart and skeletal muscles. I repeat the opinion that ion flow is not fast enough to achieve 430 volts per second.

Conclusion: **The very fast depolarisation of the spike of action potentials in conducting tissues is compatible with the hypothesis that it is achieved by electron outflow.**

CHAPTER 10

Repolarisation

It will be apparent in the four previous figures (9–12) that depolarisation (electron loss) is followed by repolarisation (restoration to the original electron density). In Figure 12, after the spike has gone from 90 to -40 mV, there is an almost immediate fast repolarisation to -5 mV. Then something quite different happens which I will deal with in the section on ion channels. Up to that point, the trace is very similar to that of an action potential of a nerve axon. We know from Weidmann's work that the cardiac electrical impedance increases, and in any case, the electron density being negative, any further electron outflow is impossible. Nevertheless, this initial increase in electron density is too fast to be accounted for by an ion flow. My suggestion is that the huge increase in electron density difference between the mitochondria at 200 mV and the cytoplasm at -40 mV produces a fast electron flow into the cytoplasm from mitochondria.

A comparison with the nerve axon spike (Figure 13) shows an even faster downstroke velocity than the 430 volts/sec of the ventricular myocardium. It is difficult to avoid the surmise that they share the same mechanism, the choice being between Na^+ inflow current and electron outflow. In the nerve axon this rapid inflow from mitochondria takes the electron density back to 70 mV (Figure 14) completing the rapid repolarisation of the nerve action potential. The mitochondria are distributed along the axons, while the heart ventricle has the most densely packed mitochondria of any organ. The duration of this nerve impulse is only one millisecond or less. The conventional

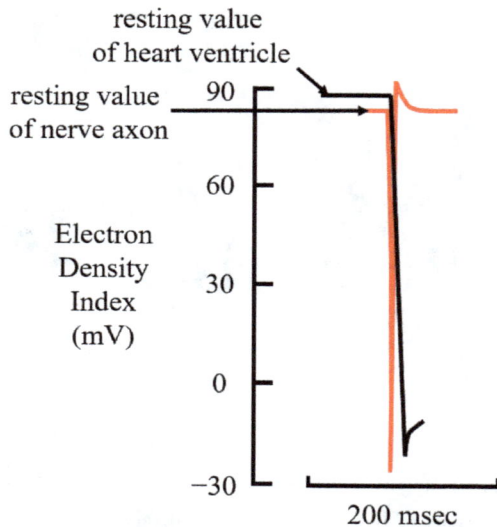

Figure 13:   Comparison of the downstroke in electron density of nerve axon and heart ventricular muscle. Not previously published.

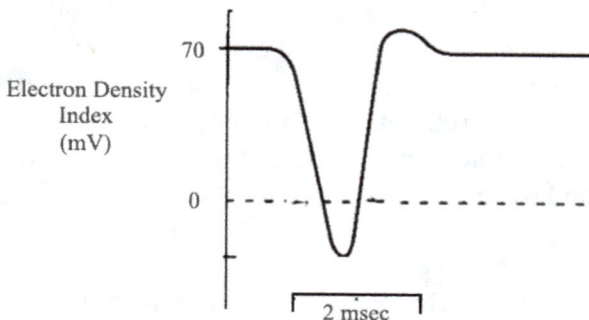

Figure 14:   Nerve axon action potential showing immediate rapid repolarisation (restoration of electron density). Not previously published.

opinion is that the depolarisation is an inflow of Na^+ and the repolarisation is an outflow of K^+. I do not accept that such ion flows are fast enough, and there is another reason reinforcing this scepticism.

Consider that nerve axons sometimes carry long trains of impulses, e.g., the train of electron density spikes illustrated in Figure 15.

If each of these action potentials causes intracellular Na^+ gain and K^+ loss, intracellular hypernatraemia and hypokalaemia results. This

Electron
Density
Index

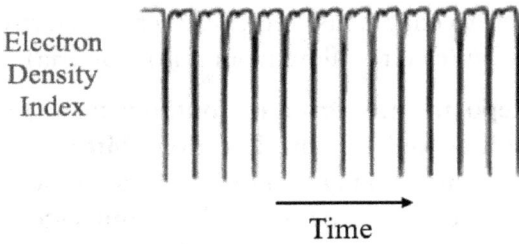

Time

Figure 15:   Train of nerve impulse in an axon, i.e., repeated sharp drops in electron density each lasting one millisecond. Not previously published.

is a clumsy mechanism compared with the electron loss and replenishment as postulated in the electron theory of electrophysiology. The reader can choose which mechanism is likely to be chosen by the process of natural selection. Fortunately, in axons, in contrast to synaptic functions, we do not have the complications associated with Ca^{2+} inflow that occur in muscles.

The reader should not confuse the question of speed of depolarisation and repolarisation with cable theory, which deals with the speed of nerve action potential conduction along the fibre. That is a set of assumptions and results relating to the propagation and interaction of electrical signals in spatially extended axons, and can explain why large diameter axons conduct action potentials faster than small diameter ones. One of the mechanisms that speed up conduction velocity in nerves with myelin sheaths is the presence of the nodes of Ranvier. These are the gaps formed between the myelin sheaths where the axons are left uncovered. Because the myelin sheath is largely composed of an insulating fatty substance, the nodes of Ranvier allow the generation of a faster electrical impulse conduction along the axon. In the electron theory the short circuit depolarisation at one node short-circuits the next, a process that is much faster than that provided by cable theory. These are exceptional; most axons are slower, some afferent and many are non-myelinated.

The action potentials of skeletal muscles, stimulated by release of the neurotransmitter acetylcholine, are similar to those in nerve axons, except that their duration is about 2.5 msec compared with 1.0 msec of axons. Skeletal muscles also show trains of action potentials, as in motor nerve neurones, and also have mitochondria distributed along the muscle fibres to enable fast replenishment of electrons to achieve fast repolarisation (electron flow from mitochondria).

Conclusion: Repolarisation from ion outflow current seems unlikely in view of resulting cellular ion depletion. Mitochondria, with the highest electron density, may be postulated to respond to increased electron density difference between them and cytoplasm to cause increased electron flow to restore cytoplasmic electron density during repolarisation.

CHAPTER 11

The calcium, sodium and potassium problems

The intracellular calcium ion concentration (Ca^{2+}) is extremely low: 10 to 100 nano molar (nM) compared with about 2 mM in the plasma — about 10,000 times lower. To me, this implies that the major function of the cell membrane (also called plasmalemma or plasma membrane, or in muscle sarcolemma) is to protect the cell contents from death due to precipitation by Ca^{2+} of the phosphate-based chemical energy system (ATP-creatine phosphate); this is called "calcium overload". But a result of natural selection is that muscle contraction depends on the action of Ca^{2+} and ATP on the protein contractile complex actomyosin. Many other cells also rely on Ca^{2+} as a second messenger to initiate intracellular metabolism and functions. Therefore, very careful means of introducing Ca^{2+} into the cell are required, just enough for the reaction to take place, mainly via the L-type Ca^{2+} ion channel. Intracellular Ca^{2+} is mopped up by the endoplasmic reticulum (sarcoplasmic reticulum in muscles). But because the cells have gained Ca^{2+} in these processes, Ca^{2+} has to be extruded in the same amount as that which entered in the first place. This is achieved by the Na^+/Ca^{2+} exchanger (NCX) that extrudes one Ca^{2+} in exchange for 3 Na^+. That the Ca^{2+} in and out must be equal in order to balance the Ca^{2+} system has been demonstrated and emphasised by Eisner (2012). (The principle is very like the secret with which an adult animal can keep a constant mass — energy out must equal food energy in!)

But the problem cannot end there, because the cells have gained "unwanted" Na^+ that needs to be extruded if one is to maintain Eisner's principle of ion flux balance for Na^+ as well as Ca^{2+}. The result is activation of the sodium pump as clearly demonstrated in the experiment of Boyett (1987).

This group increased intracellular Ca^{2+} by doubling the frequency of contractions in Purkinje fibres leading to an increase in NCX exchange and a measured (Na^+ sensitive intracellular electrode) increase in intracellular Na^+ concentration (see Figure 16, in which I have inverted the original trans-membrane trace to obtain electron density index). Immediately, with the first short interval, the active force (that reflects the amount of Ca^{2+} released internally, see top trace and Figure 17) is reduced by a mechanism called "incomplete mechanical restitution". There is a build-up of intracellular Na^+ concentration (middle trace of Figure 16).

The increased Na^+/K^+ ATPase activity caused by the increased intracellular Na^+ concentration causes an accompanying rise in diastolic electron density. Then, with continued stimulation at double the frequency, the internal Ca^{2+} and active force build up to a new steady state. This is accompanied by increase in peak electron density (top trace: lowered diastolic trans-membrane potential). The switch back to the lower stimulus frequency reverses the changes as the preparation goes exponentially back to the original steady state.

The active force signal in this and many other experiments is used as an index of internally released Ca^{2+}. Is this justified? Any attempted measurement of intracellular Ca^{2+} requires that Ca^{2+} reacts with a reagent in such a way as to produce a measurable signal. The use of active force as such a signal depends on the calibration, i.e., the relationship between Ca^{2+} as the independent variable and, in this case, active force as the dependent variable. This requires that one remove the cell membrane (so-called skinning) so that one can bathe the muscle contractile system directly in solutions of different Ca^{2+} concentrations over the required range, i.e., the range of Ca^{2+} intracellular concentrations. Kentish (1986) achieved this by dissolving off the fatty cell membrane of rat right ventricular trabeculae with detergent.

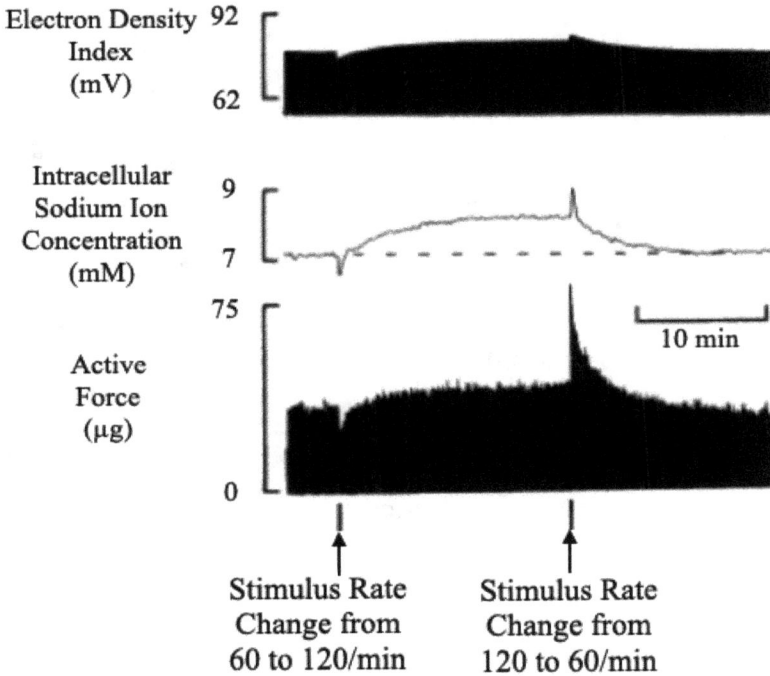

Figure 16: Doubling the frequency of Purkinje fibre excitation and contraction causes an increase in intracellular Na^+ concentration. The first beat at the arrow indicating the switch to double the stimulus rate follows the first of a series of shorter diastolic intervals. Note that the following diastolic electron density is reduced as there has been less time for Ca^{2+} to become available for release (incomplete mechanical restitution). Not previously published.

The trabeculae being a striated muscle in which the contractile apparatus consists of long chains of sarcomeres, characterised by dark bands of mostly myosin alternating with light bands of actin, it was also possible to monitor the sarcomere length using light diffraction.

The purists, not unnaturally, dislike the fact that the relationship depicted in Figure 17 is not linear, but usually, only a small part of the relationship is covered in the experiments to be described. An alternative indicator that emits light when reacting with Ca^{2+} was aequorin, reviewed by Mithofer & Mazars (2002). This has to be injected into each cell, which is much more difficult than measuring force with a simple transducer in series with the muscle. Moreover, the relationship of measured light intensity to Ca^{2+} concentration

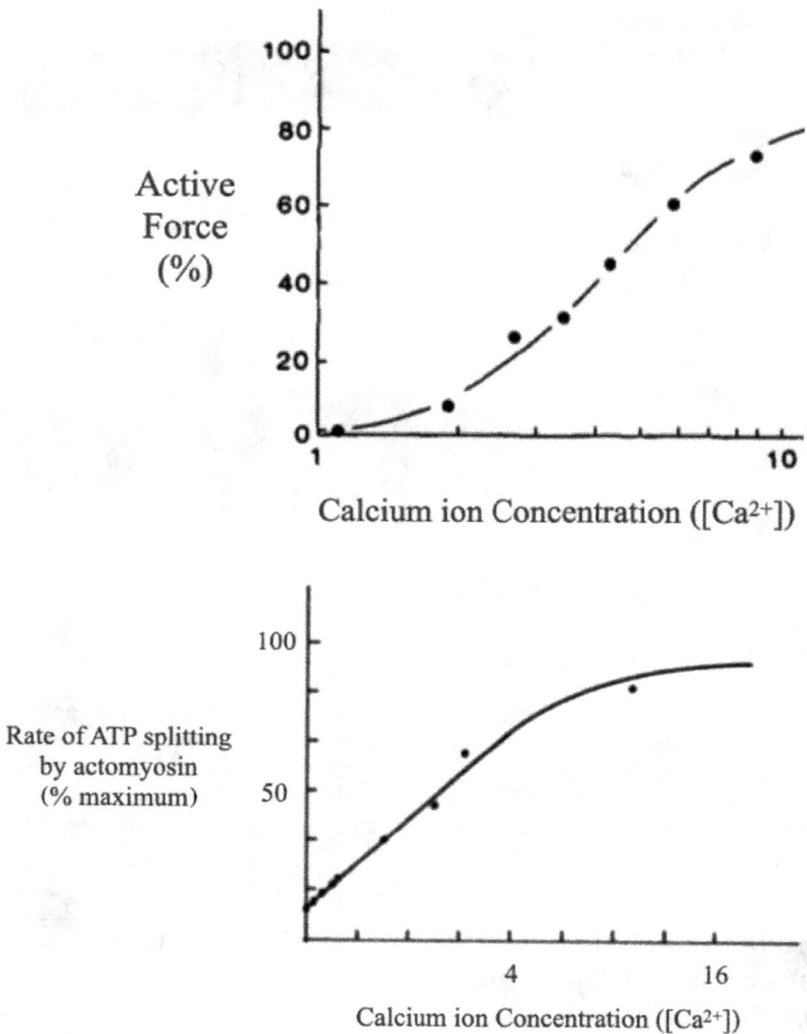

Active Force (%)

Calcium ion Concentration ([Ca²⁺])

Rate of ATP splitting by actomyosin (% maximum)

Calcium ion Concentration ([Ca²⁺])

Figure 17:   Above: The relationship between active force and intracellular levels of Ca^{2+} is not a straight line (data from the experimental records of the author). Below: Relationship, to Ca^{2+} concentration, of ATP splitting by actomyosin (biochemical reaction that produces force) obtained by Iwazumi. Reproduced with permission from Iwazumi (deceased) before his death. Note that the abscissa above is logarithmic and that below is linear. Nevertheless, the point is that both variables always go up when Ca^{2+} goes up and always go down when Ca^{2+} goes down, i.e., change in active force is a reasonable indicator of Ca^{2+} changes. Not previously published.

is not simple as:

$$[Ca^{2+}] = \{(L_0/L_{max})^{1/3} + [KTR(L_0/L_{max})^{1/3}] - 1\}/$$
$$\{KR - [KR(L_0/L_{max})^{1/3}]\} \qquad (3)$$

where:

L_0 = luminescence intensity per second;
L_{max} = total amount of luminescence present in the entire sample over the course of the experiment;
$[Ca^{2+}]$ = calculated Ca^{2+} concentration;
KR = dissociation constant for the first Ca^{2+} ion to bind;
KTR = binding constant of the second Ca^{2+} ion to bind to aequorin.

Simultaneous force and Ca^{2+} aequorin signals may be observed in cardiac ventricular muscle (Allen & Kurihara, 1980). More usefully, chemical indicators that are now available allow for intracellular Ca^{2+} detection over a very large range (<50 nM to >50 µM), and are reviewed by Paredes (2008). These agents have the advantage that one can follow the movement of Ca^{2+} within the cell. For the moment, we will use active force as an index of total cell-released Ca^{2+}.

But the calcium-sodium-potassium problem cannot end at correction of unwanted sodium by the sodium pump because the cells have gained "unwanted" K^+ that needs to be extruded if one is to maintain Eisner's principle of ion flux balance for K^+ as well as for Ca^{2+} and Na^+. K^+ channels function to conduct potassium ions down their electrochemical gradient, i.e., outward. Many such channels have been described with many different functions claimed for them, but in the present context one needs an actual measurement of K^+. This has been demonstrated in sinus node cells by Kronhaus (1978), in that extracellular K^+ concentration, close to the outer leaflet of the cell membrane, and measured with a K^+-sensitive electrode, went up and down with the action potential cycle.

Again in Figure 18, I have inverted the original trans-membrane potential trace to obtain the electron density index. The peak extracellular K^+ concentration occurred during repolarisation, but I do not share the assumption that this is the cause of all of repolarisation

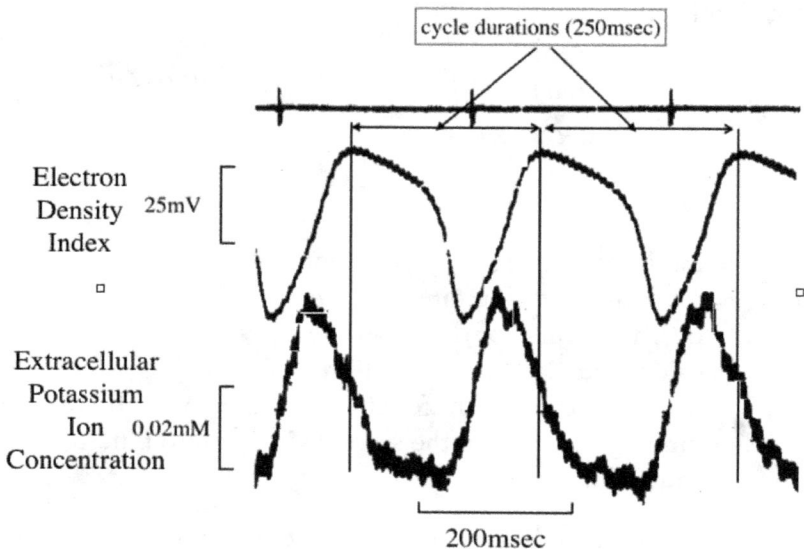

Figure 18:   Sinus node cell cycling. Upper trace shows pacemaker current loss of electrons during diastoles, triggering action potentials. Lower trace shows extrusion of potassium ions to the extracellular compartment, peaking during repolarisation. Note that the concentrations are several orders of magnitude lower than the total intracellular concentration in Figure 3. Not previously published.

(it is insufficient, and I think most of repolarisation is due to intracellular electron flow according to the electron theory).

I suggest alternatively that the average extrusion of K^+ is sufficient to achieve the K^+ flux balance necessary for the Eisner ion flux balance principle to apply. Unfortunately, although intracellular recordings of K^+ concentration have been made in other cells using K^+-sensitive electrodes, possible oscillations during cycling in sinus node cells have not been performed to my knowledge. It is not clear to me through which mechanism the K^+ passes out. Internal alkalinity stimulates K^+ outflow, but only at unphysiological levels of pH. Outward permeability takes about 40 minutes, whereas the K^+ extrusion (Figure 18) occurs in about 100 msec. Excluding these, the answer would seem to be the operation of one or more K^+ channels. Upon changes in trans-membrane potential, these channels open and allow passive flow of K^+ ions from the cell, but how the quantity is matched to the "unwanted" K^+ from sodium pump activity is not clear. K^+ channels have trans-membrane helices spanning the lipid bilayer. There are three major classes but studies of these do not define the quantity of ion passing out compared with K^+ passing in through the sodium pump. There are at least 7 types of K^+ channel blocker, which is one reason for me not liking to extrapolate from pharmacology to physiology! The one I have used in patients is amiodarone which lengthens the cardiac ventricular action potential (and QT interval of the ECG). I postulate that the drug is preventing some K^+ extrusion, thus increasing intracellular K^+ concentration that adds more positive charge to the cytoplasm and therefore delays repolarisation (more electrons needed from the mitochondria).

[There is a considerable amount of literature on molecules called K^+ ATP channels and K^+ channels, which mostly consist of molecular biological and pharmacological data and speculation on their roles in modulating electrical activity. According to my reading of the literature, there is a lack of study of their role in intact cells and tissues, as is also the case for a multitude of other K^+ channels that have been described. What exactly is the mechanism of K^+ extrusion demonstrated in sinus node cells, for which I postulate the necessary K^+ ionic balance? Colin Nichol's statement (2006),"Many questions remain to be addressed, but the cloning of constituent subunits of the K^+ ATP channel and the crystallisation of structurally related proteins have provided the raw materials to gain a detailed understanding of how nucleotides interact with the K^+ ATP channel and an explanation of the molecular basis of channel activity and of channel-dependent human disease."; this does not help me puzzle out my vital question (maintenance of K^+ ionic balance)].

These three problems (the calcium, sodium and potassium problems) have effects on the electrical recordings from cells; that is why they should be understood in electrophysiology:

1. Ca^{2+} entering through the Ca^{2+} channel involves a cellular loss of 2 electrons per Ca^{2+}, which has 2 electrons missing from its valence orbital. This results in further depolarisation on top of that due to electron outflow.
2. $3 Na^+$ per Ca^{2+} extruded, entering through NCX, involves cellular loss of one electron per Na^+ which has one electron missing from

its valence orbital. This results in further depolarisation on top of that due to electron outflow and Ca^{2+} inflow.

3. 3 Na^+ exchanging for 2 K^+ in the sodium pump involves a cellular gain of one electron per exchange, explaining the observed decrease in diastolic trans-membrane potential (increased electron density).

4. K^+ exiting the cells during repolarisation involves a cellular gain of one electron per K^+, thus enhancing the cellular gain of electrons from mitochondria.

Principle: ionic balance, i.e., equal ionic entry and exit from cells, is necessary for a steady state. A change of activity involves a transient perturbation until a new steady state is reached.

CHAPTER 12

Excitation-contraction coupling in muscles

Depolarisation in muscles triggers contraction. This process is called excitation-contraction coupling. The major reason this can happen is the presence of the organelle endoplasmic reticulum (sarcoplasmic reticulum (SR) in muscle). This acts as an internal store that can release Ca^{2+} for contraction and take it up again. However, this has effects on electrical events which vary with muscle type.

The effect of the Ca^{2+} problem on skeletal muscle electrophysiology

Skeletal muscle responds to a single action potential with a twitch contraction. Trains of action potentials in skeletal muscle allow control of the duration of contraction, allowing voluntary control by the organism via the central nervous system and the motor neurones. The excitation-contraction coupling in skeletal muscle has been summarised by Calderon (*2014*). I quote, "The term excitation-contraction coupling describes the rapid communication between electrical events occurring in the plasma membrane of skeletal muscle fibres and Ca^{2+} release from the SR, which leads to contraction."

The sequence of events in twitch skeletal muscle involves:

(1) initiation and propagation of an action potential along the plasma membrane;

(2) spread of the potential throughout the transverse tubule system (T-tubule system);

(3) dihydropyridine receptor-mediated detection of changes in membrane potential;

(4) allosteric interaction between these and SR ryanodine receptors;

(5) release of Ca^{2+} from the SR and transient increase of Ca^{2+} concentration in the myoplasm;

(6) activation of the myoplasmic Ca^{2+} buffering system and the contractile apparatus; followed by

(7) Ca^{2+} disappearance from the myoplasm mediated mainly by its re-uptake by the SR through the SR Ca^{2+} adenosine triphosphatase (SERCA), and under several conditions, movement to the mitochondria and extrusion by the Na^+/Ca^{2+} exchanger (NCX).

In relation to electron theory, we substitute the term electron density for membrane potential.

As illustrated for Purkinje fibres (Figure 16), an electrophysiological consequence of NCX is that 3 Na^+ enter the cell in exchange for 1 Ca^{2+} expelled. The net effect of 3 positive charges entering, as these are atoms lacking an electron in their valence orbital, in exchange for 2 positive charges, is loss of one electron per turn to the cytoplasm, i.e., more depolarisation (reduced electron density). This requires increased electron production by the mitochondria. The sodium pump (Na^+/K^+ ATPase) responds to the increase in intracellular Na^+ concentration caused by NCX activation. This is an exchange of 3 Na^+ out for 2 K^+ in, so by using the same reasoning in reverse, one concludes that there is increased electron density as a result. This is more easily demonstrated in cardiac muscle in which these adjustments are also required (Figure 16).

The removal of some Ca^{2+} by NCX inevitably causes some depletion of intracellular Ca^{2+}, but this is made up for by an equal gain of Ca^{2+} through the Ca^{2+} ion channel. In skeletal muscle cells, voltage-gated Ca^{2+} channels in the transverse tubule membranes interact directly with ryanodine-sensitive Ca^{2+} release channels. There is also Ca^{2+} binding and release from the cell membrane. Living cells maintain a huge trans-membrane electric field across their membranes. This electric field exerts an effect on the membrane because the membrane surfaces are highly charged. Electric field force equals trans-membrane electric potential divided by membrane thickness which is in the range of 5–8 nanometres (nm). For a plasmalemmal thickness of 5 nm, the resulting field force is about 100,000 volts per centimetre (Alberts, 2002). It has been observed that calcium is bound to the inner leaflet of the plasmalemma in polarised cells, presumably as a result of the electric field force and the negative charge of the anionic phospholipids. In the context of calcium binding, phosphatidylserine is of particular interest, as it has pairs of extra electrons ready to bind Ca^{2+}, which has two electrons missing from the calcium atom's valence orbital. Langer postulated that the binding was dependent on the strength of the electric field force being sufficient to exclude charged particles from the interlamellar space of the cell membrane, even protons (hydrogen ions). Therefore acidification (adding protons) reduces the binding while alkalinisation (removing protons) increases the binding (Langer, 1985). Loss of trans-membrane electric potential, by allowing membrane protons to increase, causes loss of Ca^{2+} binding, and I suggest that this occurs at electron density index of 40 mV in sinus node cells, resulting in the action potential, which is associated with the onset of Ca^{2+} current (Hagawara, 1988).

CHAPTER 13

The effect of the Ca^{2+} problem on cardiac muscle electrophysiology

Excitation-contraction coupling in ventricular muscle differs from that in skeletal muscle in that the spike that opens the Ca^{2+} channel causes a sustained inward current (causing the downward electron density bulge in Figure 12, that follows the initial fast depolarisation). The action potential is thus prolonged. The rise in intracellular Ca^{2+} concentration activates the ryanodine receptors on the terminal cisternae of the SR, which then releases a large amount of calcium to react, plus ATP, with the actomyosin system to produce the contraction of the vital heartbeats. Study of cardiac excitation-contraction coupling is more rewarding than the fundamentally similar underlying mechanism in skeletal muscle because the heart performs regular contraction/pause/contraction/pause/...... and does not follow the signals from a nerve in order to contract. It goes on working when there are no nerves to it, e.g., in cardiac transplantation.

In isolated strips of ventricular muscle (usually papillary muscle or trabeculae), Ca^{2+} release is represented by its effect on the upstroke of the isometric force curve in Figure 19 (see Figure 17 and associated discussion for justification).

This high (but not dangerously high!) intracellular Ca^{2+} concentration initiates Na^+/Ca^{2+} (NCX) exchange which further prolongs

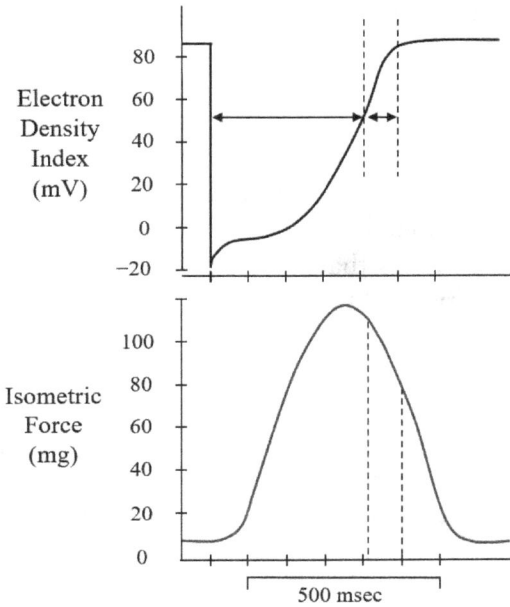

Figure 19: Action potential and force response in cardiac ventricular muscle. Not previously published.

the plateau of the action potential (upper trace in Figure 19). During this time, it is not possible to elicit another action potential because the cytoplasm is depleted of electrons. This, as indicated by up to the left dotted line in the upper trace, is called the "absolute refractory period". It is difficult to understand how the conventional theory (depolarisation by inward Na^+ current) can explain the absolute refractory period. After all, there is still plenty of Na^+ in the extracellular fluid and a very large concentration gradient for Na^+ from the outside to the inside of the cell. Why should there be no Na^+ current? By contrast, electron density is negative, i.e., the cell is depleted of electrons, the electrical impedance is high (Figure 11) and therefore, continued outward electron flow is not possible, so further depolarisation and action potential do not occur.

This is then followed by a period between the vertical dotted lines when small action potentials can be evoked, getting larger with time. The period is called "relative refractory period" and the growth of the action potential signal with time is called "electrical restitution".

This is **not** the cause of mechanical restitution (that is the delay before uptake of Ca^{2+} by the SR to produce relaxation becomes available for re-release). Eventually, when Ca^{2+} is taken back into the SR internal store (indicated by the fall of isometric force, i.e., relaxation in the lower trace of Figure 19), the concentration of Na^+ (from NCX) is reduced by the sodium pump, and the continued flow of electrons from mitochondria completes repolarisation.

The proposal implies that the long action potential (duration of low electron density) of cardiac ventricular muscle could be due to the number of intracellular cations (atoms with missing electrons) slowing the repolarising effect of the electron flow from mitochondria. This idea can be tested by adding even more "positive" charges to the cytoplasm by inducing intracellular acidosis, i.e., adding protons. The result is indeed a further prolongation of the action potential reported consistently by many authors.

CHAPTER 14

Mechanical restitution and the optimal contractile response

Edman and Johannsson (*1976*), studying rabbit papillary (cardiac ventricular) muscle, inserted stimuli at a series of test pulse intervals. These were applied after a train of regular stimulus-induced contractions until a steady state was established. The amplitudes of the active forces elicited by these test stimuli were plotted against the test pulse intervals. Stimuli at short intervals following the steady state train produced weak contractions (Figure 20). As the durations of test pulse intervals were lengthened, the amplitudes of force development elicited became greater and greater to reach a maximum. The authors called this maximum the optimal contractile response.

With the assumption justified in Figure 17 and the associated discussion, we can substitute [Ca^{2+}] for force amplitude on the ordinate.

Why is so little Ca^{2+} released after a short test interval?, i.e., what is the situation as regards Ca^{2+}? There has just been a contraction caused by the reaction between released Ca^{2+}, ATP and actomyosin. As in the case of skeletal muscle described above, Ca^{2+} disappearance from the myoplasm is mediated mainly by its re-uptake by the SR through the SR Ca^{2+} adenosine triphosphatase (SERCA). This causes relaxation (down-going red trace in Figure 19). A stimulus during the relative refractory period is predicted to produce more Ca^{2+} release

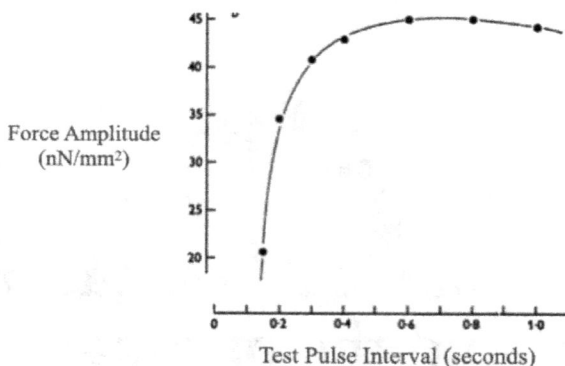

Force Amplitude
(nN/mm²)

Test Pulse Interval (seconds)

Figure 20: Amplitude of the isometric test twitch in relation to the test pulse interval after a 2.5 second priming period. (Data from the experimental records of the author.) Not previously published.

before the relaxation is complete. Actually this does not occur. Once mechanical restitution begins, as one increases the duration of the test pulse interval, more and more Ca^{2+} will become available for re-release, thus causing increasing amplitude of force responses. This process is called mechanical restitution but actually means restitution of releasable Ca^{2+}.

The conclusion is that it takes about 800 milliseconds for all the Ca^{2+} in the SR to become releasable. This led Wohlfart (*1979*) to conceive a two-compartment model of the SR. There was an uptake compartment accumulating Ca^{2+} during relaxation and a release compartment to which Ca^{2+} from the uptake compartment was transferred over a time course of 0.8 seconds. When the total circulating Ca^{2+} was increased by increasing the frequency of stimuli in the priming period and then introducing the test interval protocol, there was an increase in the overall force versus test interval curve (Figure 21).

The results from Edman and Johannsson's (*1976*) study show that the time course of mechanical restitution remains the same under these circumstances, but the increasing height of the curves reflects the greater amount of Ca^{2+} induced by the increased priming stimulus frequency. This increase in circulating Ca^{2+} is also shown in Figure 16 when stimulus frequency was doubled but was grossly underestimated because with the first interval after discontinuation of the 120/minute period, there was a much larger active force on that beat, i.e., after

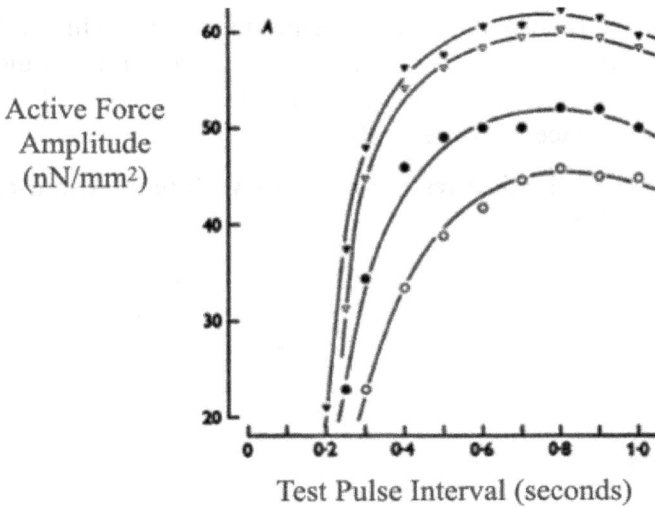

Figure 21:   Amplitude of the isometric test twitch in relation to the test pulse interval after priming periods of increasing stimulus frequency per second from runs 1 to 5: open circles runs 1 and 2, closed circles 3, open triangles 4, closed triangles 5. (Data from the experimental records of the author.) Not previously published.

the first one-second interval (close to the optimal contractile response interval).

Principle: **The total Ca^{2+} circulating in cardiac muscle is not indicated by the changes in steady state level of force, but only after inserting a test pulse interval of 0.8–1.0 second.**

This explains the high initial force (and Ca^{2+} release) with the first long interval in Figure 16 called "post-tachycardia potentiation". Other evidence of the increasing circulating Ca^{2+} with increasing stimulus rate is provided by the increasing amplitude of the aequorin signal (Allen & Blinks, *1978*).

What remains of this part of the discussion is the question, "At what moment does mechanical restitution begin?" This was answered by Arlock (*1981*) who used ferret papillary muscle in a sucrose gap preparation. This set-up allows one to replace the normally elicited action potential (Figure 12) with a square wave of low electron density. The priming period consisted of a train of 60/minute square wave depolarisations of 200 msec duration. There were then two protocols for comparison. In the first protocol, mechanical restitution was determined in the usual way by varying the test pulse interval. In the second protocol, the last depolarisation before the test pulse intervals was prolonged to 500 msec. The resulting mechanical restitution curve was displaced in time by 300 msec.

Conclusion: **mechanical restitution starts with repolarisation.**

CHAPTER 15

Internal calcium ion release and recirculation

In 1975, Fabiato & Fabiato established, using physically skinned cardiac cells, that Ca^{2+} was released from the SR near the end of the sarcomeres (terminal SR or terminal cysternae) by Ca^{2+} acting upon their ryanodine receptors, i.e., calcium-induced calcium release. With the advent of diffusible fluorescent Ca^{2+} indicators it was possible to show that this released Ca^{2+} spread to the rest of the sarcomeres, so that the highest concentration at first was at the ends and the lowest in the centre of the sarcomeres. This was also found by Picht (*2011*) using modern Ca^{2+} detection. His group also found that after release the Ca^{2+} concentration in the longitudinal SR wrapped around the contractile filaments was no different from that in the terminal SR. This is to be expected, because the whole of the SR may have recently emptied its Ca^{2+} into the cytoplasm to produce contraction. The study does not tell us how Ca^{2+} taken up by the SERCA mechanism to produce relaxation travels from the longitudinal SR to the terminal SR in a time span of up to 800 msec. This is much slower than the diffusion following release, reaching peak concentration as soon as approximately 30 msec according to fluorescent Ca^{2+} indicators, e.g., Picht's study. It also leaves us with no explanation of why the movement does not begin until repolarisation is complete. Does the movement require that the electron density in the cytoplasm be maximal? According to Sanchez (*2018*), in skeletal muscle, release of Ca^{2+} into the cytoplasm is balanced by concomitant counter-ion currents across the SR membrane, because there is no voltage change, interpreted here as no electron density change within the SR; note that in Figure 8 we recorded an electron density of zero for endoplasmic reticulum. Restoration of a cytoplasmic electron density to 85 mV upon repolarisation would create a large potential difference between the cytoplasm and the SR, which might be relevant in the starting of the Ca^{2+} restitution process?

The rate of Ca^{2+} release has been estimated as reaching a peak at the sarcomere end in about 3 msec (Holash & Macintosh, 2018). Figure 22 depicts my interpretation. It is extraordinary that I had to correct the labelling by other authors of cytosol to cytoplasm. One wonders that persons in 2011 had presumably never heard of MRI which has been around since 1973 and is now a regularly used imaging technique in medical diagnosis. MRI does not work if the body's cells are full of liquid, but only if they are full of gel. There is no cytosol. The model does not have a time scale, so it would not have been possible to include the timing of mechanical (Ca^{2+}) restitution and the wait for

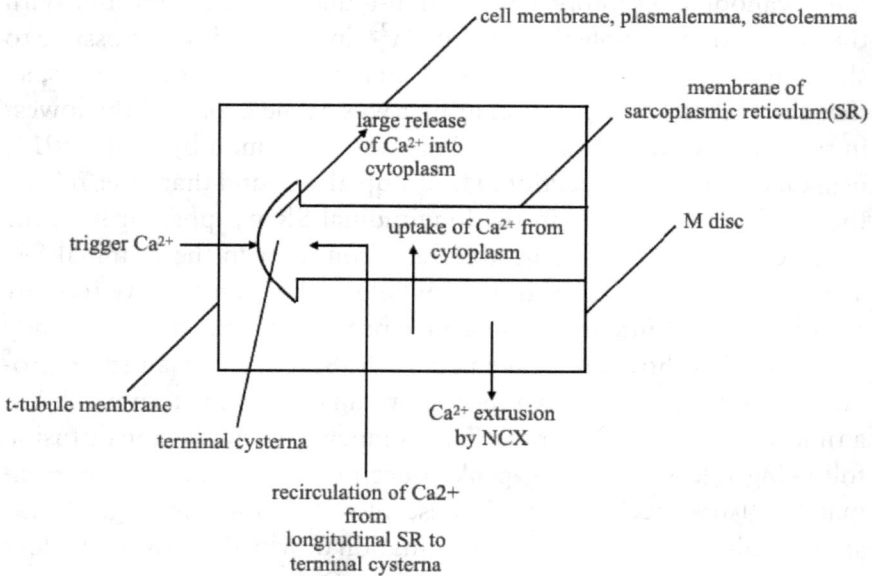

Figure 22:   Model of proposed uptake compartment (uptake of Ca^{2+} by SERCA of longitudinal SR), release compartment (terminal cysterna) and Ca^{2+} recirculation. Not previously published.

mechanical restitution to start until the cytoplasmic electron density recovered to 85 mV. However, the model is correct in showing that a proportion of Ca^{2+} recirculates from SR uptake to the release site at the terminal cysternae. The exit of the remaining Ca^{2+} is indicated in the lower right-hand part of the sarcolemma; this will exit the cell through the NCX. The next question is, "How much recirculates and how much exits?"

In Figure 16, one can observe that on the first stimulus after a one-second interval when the stimulus rate is returned to 60/minute, the peak force is much enhanced (post-frequency potentiation). The following recovery follows an exponential time course. Analysis of this time course allows one to calculate the recirculation fraction (ter Keurs, 1990). They found that the contractility of a beat (reflecting peak $[Ca^{2+}]$) depended on

$$[Ca^{2+}]_n = RF \; x[Ca^{2+}]_{n-1} + EAPD \; x(APD_{n-1} - D) \qquad (4)$$

where:

$[Ca^{2+}]_n$ = peak-released calcium of the beat;

RF = recirculated fraction of calcium;

$[Ca^{2+}]_{n-1}$ = peak-released calcium of the previous beat;

$EAPD$ = a constant of proportionality between APD_{n-1} and $[Ca^{2+}]_n$;

APD_{n-1} = action potential duration of the previous beat;

D = a constant with dimensions of time, postulated to be a 'dead time' within the action potential not contributing to calcium entry.

Thus, a proportion of the Ca^{2+} that is released on a beat comes from Ca^{2+} re-circulated from the previous beat, and in addition, some come from calcium entering (via the Ca^{2+} channel) from the previous beat depolarisation action potential.

Principle: Ca^{2+} released on a beat to cause proportional contraction does not come from the Ca^{2+} entering the cell on that beat, but only from the previous beat via recirculation and the Ca^{2+} entry of the previous beat.

The Ca^{2+} entry on the beat itself causes the release of the stored Ca^{2+} (stored from the previous beat) by the calcium-induced calcium release

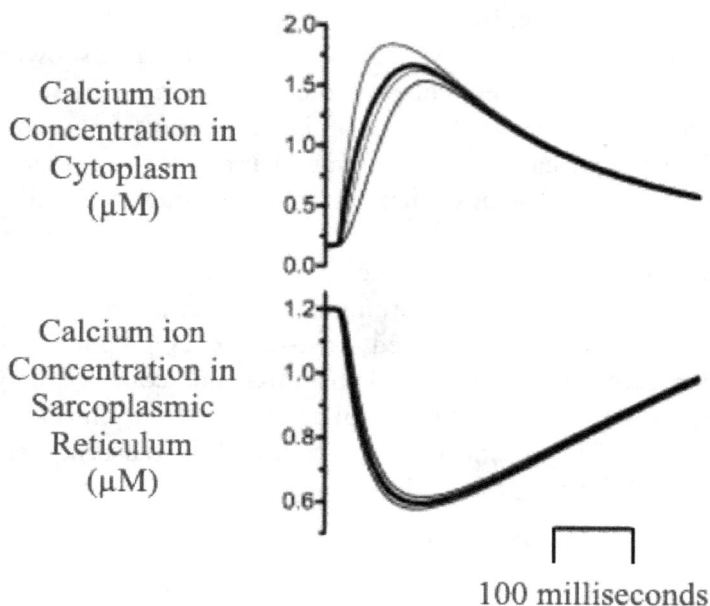

Figure 23:   Time dependence of Ca^{2+} concentration (fluorescent Ca^{2+} indicator) upon release from the terminal cisternae (data derived from Picht, 2011). The Ca^{2+} release into the cytoplasm (upper trace) comes from that already stored in the SR (lower trace); this empties with a similar time course to the cytoplasmic gain curve. The differently coloured curves were measurements from different locations along the sarcomere. Not previously published.

mechanism (Fabiato & Fabiato, *1975*), i.e., trigger calcium acting though the ryanodine receptors of the terminal cisternae.

The awkwardness of the need for D, the constant with dimensions of time, postulated to be a 'dead time' within the action potential not contributing to calcium entry, can now be explained with knowledge from more recent data using fluorescent Ca^{2+} indicators.

It is clear from the time curves in Figure 23 that the Ca^{2+} concentration is not constant during the action potential, and, as far as Equation (4) is concerned, not simply linearly proportional to APD_{n-1}.

Let us not forget that the depolarisation of APD_{n-1} at 430 volts/sec (Berecki, *2010*) requires the reader to choose between the conventional inward current carried by sodium ions, or, what in my opinion is more feasible, an outflow of electrons in the one-way system postulated above.

There is a simpler way of quantifying the recirculation fraction, which has the advantage that it can be applied to intact animals and humans as long as one uses the maximum rate of rise of left ventricular pressure as an index of intracellular $[Ca^{2+}]$ because peak force cannot be measured in ejecting hearts. (The relationship in Figure 17 requires that the muscle be held at a constant length, i.e., isometric.)

The protocol is similar to that of Boyett (*1987*), except that only the first two potentiated beats after the change back to 60 beats per minute are required, but a range of levels of potentiation is required, i.e., the priming period is different for each run. One can use different frequencies when using post-frequency potentiation (Boyett, *1987*), different pre-ectopic (extra-systolic) intervals when using post-ectopic (extra-systolic) potentiation (Philips, *1990*), different priming intervals when using post-paired pacing potentiation (Mansfield & McDonald, *1965*), or different voltage clamp durations when using post-voltage clamp potentiation (Arlock, *1992*).

We go back to the simpler scheme of Wohlfart (Figure 24).

Here again, it is postulated that the release of Ca^{2+} comes from the Ca^{2+} that were stored in the SR, some that were not extruded on the previous beat and some that were taken up from the extracellular compartment on the previous beat.

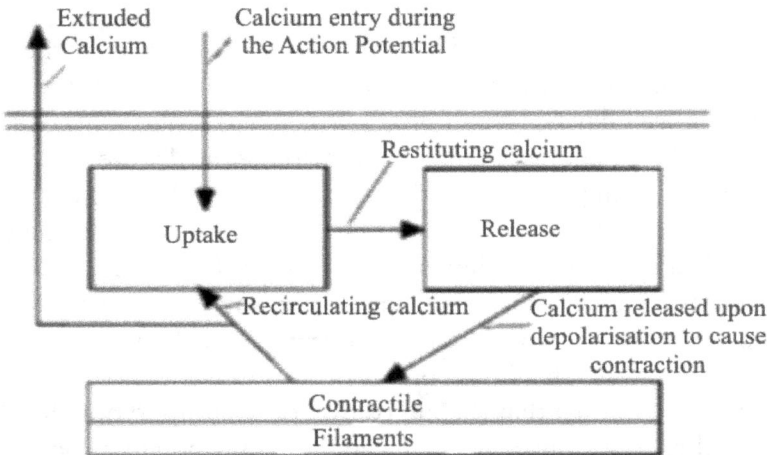

Figure 24:   Diagrammatic representation by the author of the calcium recirculation hypothesis of Wohlfart. Not previously published.

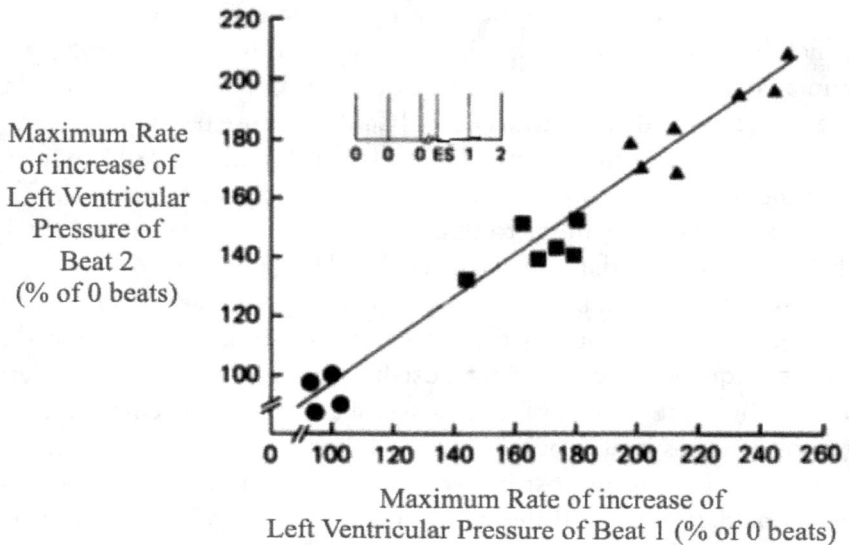

Figure 25: Data obtained from one of the author's experiments in collaboration with others. The inset indicates the stimulus protocol. Varying the interval before the extrasystole (ES) produces varying contractility on beat 1 which decays to less potentiation on beat 2. A plot of contractility on beat 2 (ordinate) against that on beat 1 yields a straight line by linear regression analysis, the slope of which in this case is 0.75, i.e., 75% of the Ca^{2+} released on beat 1 recirculates to beat 2. Not previously published.

To obtain the recirculated fraction, one needs to measure the influence of the previous beat, which is most easily done in a state of potentiated total cellular Ca^{2+} and plotting the contractility of the second potentiated beat against that of the previous beat (Figure 25).

This was a simpler method than the exponential fitting performed by ter Keurs, but it is reassuring that both methods yield the same result when compared within each experiment (ter Keurs, *1990*).

Further insights into "dead time" were sought using the voltage clamp technique. A strip of ventricular muscle is held within a sucrose gap apparatus, which allows control of the intracellular electron density. It is well known that the imposition of what up to now has been described as a prolonged voltage clamp, causes potentiation of subsequent beats and is the basis of producing increased contractility by paired pacing.

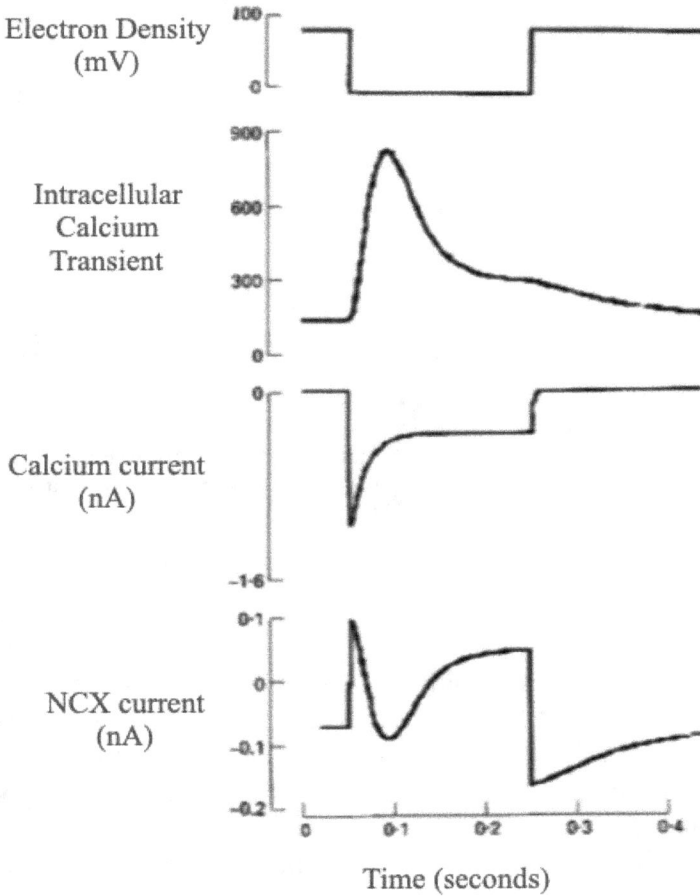

Figure 26:   A figure to show changes in electron density (ferret ventricular muscle). Emptying of the cytoplasmic electrons artificially for a longer duration than that of a normal action potential (top trace) causes prolonged opening of the Ca^{2+} channel (3rd trace), a prolonged occupation in the cytoplasm with Ca^{2+} (2nd trace), and a resurgence of NCX activity (bottom trace). Not previously published.

It is then possible to determine what is happening in this unphysiological situation. One is able to observe that the extended period of depolarisation, over and above normal action potential duration (Arlock, *1981, 1992*), causes a continued release to the cytoplasm of Ca^{2+} (Harrison, *1989*), requiring further activation of the NCX.

The effect of the extra Ca^{2+} on the subsequent normally stimulated beat is illustrated in Figure 27. It takes about 100 milliseconds before

Peak Force of
the beat one second after
a voltage clamp
to reduce electron
density to zero
(% of steady state force)

800

600

400

200

20   100   1000   2580

Clamp duration (milliseconds)

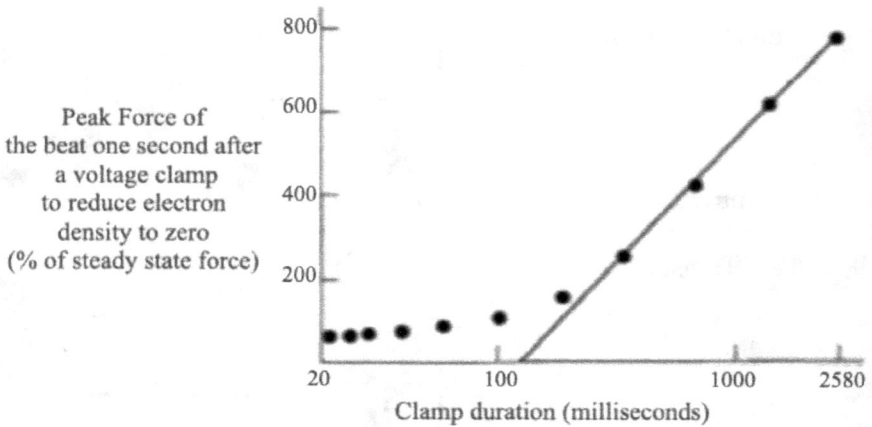

Figure 27:   Data obtained in one of the author's experiments. Relationship of decay of post-clamp contractility to clamp duration in experiments as illustrated in Figure 26 with varying clamp duration runs. Not previously published.

the full post-voltage clamp potentiation is apparent. Curiously, there is very little potentiation of the post-clamp beat when the clamp is only about 100 milliseconds in duration. Could this be an explanation for the "dead time" in the recirculation equation? It is noteworthy that there is an initial spike of NCX current (Figure 26) at the time of the initial spike of the calcium transient and little of that calcium goes into the recirculation process. It may also be responsible for the start of mechanical restitution not occurring until just after 100 milliseconds (Figure 20).

Is there any limit to the amount of calcium that the SR can take up? This question was explored by Noble (2006). The experiment illustrated in Figure 26 indicated that with long duration clamps, Ca^{2+} continued to flow into the cells, which showed some continued low force (not shown). Decay curves are normally analysed using a mono-exponential decay function, which assumes that a fixed fraction of activator calcium ions is recirculated from one beat to the next. It was postulated that there might be deviations from such a mono-exponential expression at high levels of contractility. In single sucrose-gap voltage clamp experiments of isolated ferret papillary muscle, very high contractility was obtained by potentiation due to prolonged clamp depolarisations. A bi-exponential decay was obtained in which the initial decay is much faster than the subsequent slower decay, as judged by residual variance of least-squares exponential fitting and by analysis of covariance using a linear equation (force of beat versus force of previous beat). In the slower decay period (physiological range), the decay was identical to that following post-extrasystolic potentiation in the same muscles studied with conventional stimulation.

Fact: At unphysiogically high levels of intracellular Ca^{2+}, there are 2 processes (fast then slower) contributing to the Ca^{2+} that is released

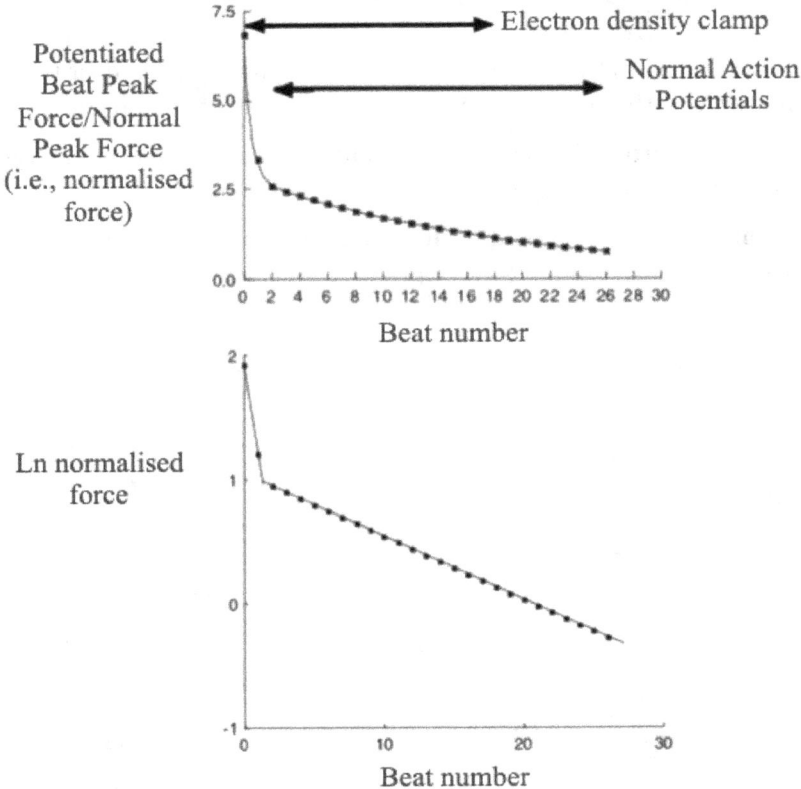

Figure 28: Data obtained in one of the author's experiments. Plots of beats after a series of one-second intervals after a voltage clamp to electron density $-20\,mV$. Upper plot: linear ordinate suggesting two exponential decays. Lower plot: logarithmic ordinate confirming two exponential decays, a fast one at very high potentiation, and a slower one in the more normal range of contractility. Not previously published.

on the subsequent electric depolarisation after the time for optimal contractile response.

I am not sure of what this means, but I suggest that there is a maximum quantity of Ca^{2+} that the sarcoplasmic reticulum can take up. When that quantity is exceeded, incoming Ca^{2+} goes straight to the contractile filaments, if my interpretation of Figure 28 is correct. Perhaps a more convincing demonstration of this phenomenon is obtained with the original recirculation plot of second potentiated beat force against first potentiated beat force, each preceded by one-second intervals (Figure 29).

For a single exponential decay between the first two potentiated beats, a single straight line should be obtained (see above). This experiment (Figure 29) yields two distinct linear phases instead of the one compatible with mono-exponential decay of potentiation.

Conclusion: From the descriptions and discussion concerning vertebrate animals including humans,

1. From theoretical considerations, depolarisation of living cells by sodium ions, coupled to repolarisation by potassium ions, is

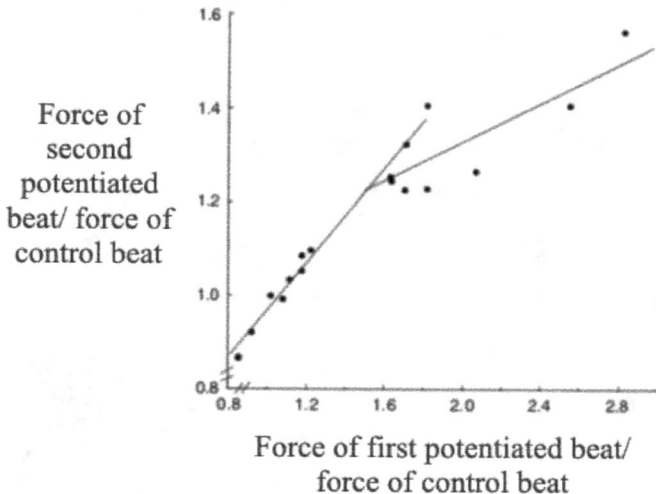

Figure 29:   Original figure, plotting data obtained in one of the author's experiments. Varying the extra-systolic interval for post-extrasystolic potentiation beats. This is the plot of contractile force of the next beat (ordinate) against that of the first potentiated beat (abscissa). Not previously published.

rejected in favour of depolarisation by electron outflow and repolarisation by electron inflow from mitochondria.

2. Data are compatible with the idea that depolarisation opens a calcium ion channel, and that that causes, and is reinforced by, calcium release from endoplasmic reticulum, which, in turn, causes the functional reactions with cell components.

| CHAPTER 16 |

Objections to the non-electromagnetic theory of striated muscle

Prior to the discovery and confirmation of the sliding filament mechanism of contraction (next paragraph), the theory of contraction was dominated by the idea of conformational change in long macro-molecules, based on the assumption of coiling of molecules from an extended state that manifests itself as muscular contraction.

In Figure 30 the agreed basic structure shows the thick myosin filaments and the inter-digitating thin actin filaments. When the sarcomere is activated by Ca^{2+} and ATP, the thin filaments slide into the gaps between the thick filaments, narrowing the light bands next to the Z disks while the dark bands, where the thick filaments are situated, do not narrow. This is the sliding filament theory of muscle contraction. Further details of the structural arrangement are provided in Figures 31 and 32.

Where there is dispute is whether the generation of the force on the thin filaments is mechanical or electromagnetic.

The heads that stick out from the myosin filaments consist of heavy meromyosin, which is the part with enzymatic activity, splitting ATP under the influence of Ca^{2+}. Illustration of the structure is difficult because of its three-dimensional character. The rods are arranged back-to-back on either side of the M disc, so that there is an area

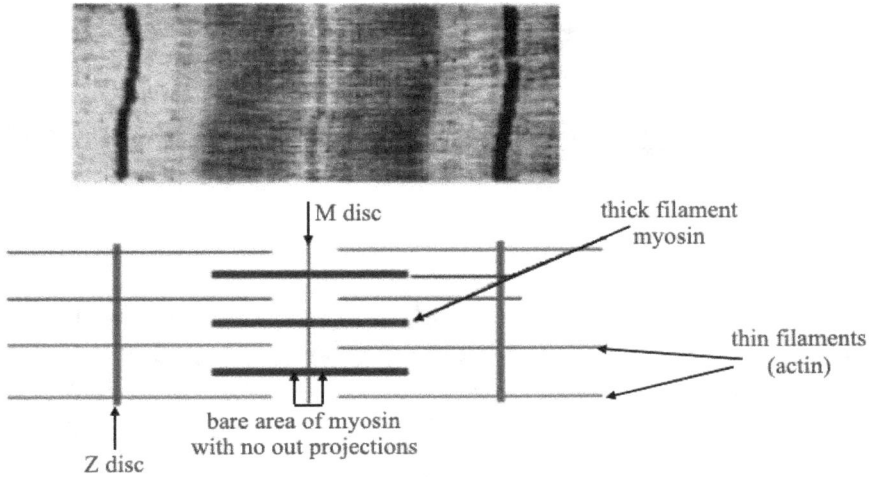

Figure 30: Structure of striated (skeletal and cardiac) muscle. Above microscopic photograph, below diagrammatic representation. Thick filaments consisting of poly-merised myosin in the centre of the sarcomere are 1.6 μm long joined in the centre by the M disc. Thin filaments consisting mainly of polymerised actin 1.0 μm long are attached to the Z discs and inter-digitate with the thick filaments. Source: Wikimedia Commons, public domain.

bare of heads. The heads all point towards the sarcomere ends. Each filament twists helically. On one strand a head sticks out every 43 nm, but the adjacent strand is displaced and the head sticks out at a 60 degree angle, and the next at another 60 degrees, and the next at another 60 degrees brings them into the same plane as the first but opposite at 180 degrees. The longitudinal shift is 43 nm so that a location on the thin filament passes a head on the myosin filament every 14.3 nm. Each unit of the helical actin filament is 38 nm.

I would hope that it is apparent from the cartoon (Figure 33) that this difference in the intervals between active sites on the two sets of filaments (called the Vernier effect) simply cannot work. It is worse than that. The theory proposes that the oars are not rigid but have the capacity to shorten to pull the thin filaments along — like nano-muscles with conformational change in nano-molecules, based on the assumption of coiling of the molecules from an extended state that manifests itself as muscular contraction. This sounds very like the pre-sliding theory described above for the whole muscle!

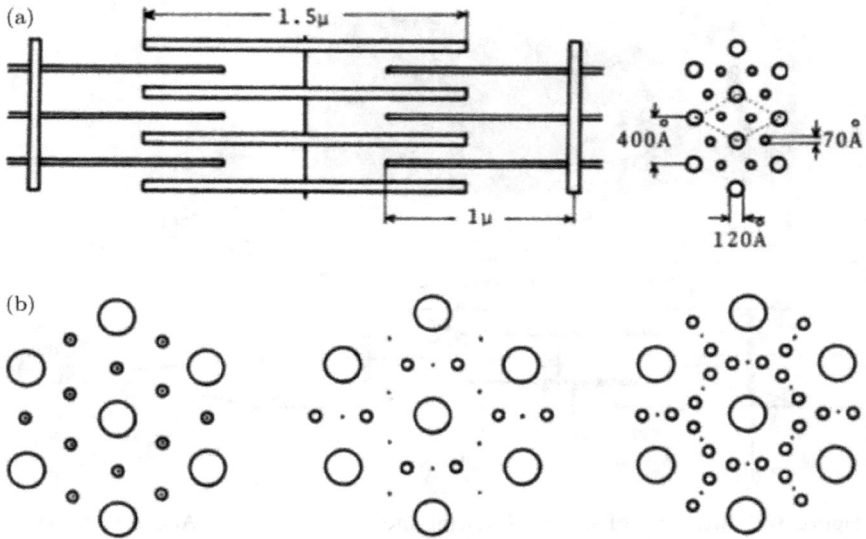

Figure 31: (a) Longitudinal arrangement of filaments. A schematic sarcomere and the constituent thick and thin filaments. They slide past each other as sarcomere length varies. To the right one observes the arrangement in transverse space where the filaments overlap. Å = angstrom. (b) Transverse arrangements. (Left) Stable positions of the thin filaments in insect flight muscle. They occupy intrinsic singular lines (small dots), hence there are three thin filaments in a unit. (Centre) Stable positions of the thin filaments in vertebrate skeletal and cardiac muscle. The thin filaments are paired up across a line where there was an intrinsic singular line. There are two thin filaments in a unit cell. (Right) Stable position of the thin filaments in a type of insect and smooth muscle with the maximum number of pairs, six filaments in a single unit. Reproduced with permission from Iwazumi (deceased) before his death.

However, if you pull at an angle, the parallelogram of forces dictates that there is a force going across that will pull the thin filaments into the middle. i.e., there is lateral instability. If you have a tug-of-war game, make sure that your team is in a straight line with flag and the course. Any force exerted that contributes to the team getting out of line will be wasted and will contribute to loss of the game.

Another difficulty in accepting the non-electromagnetic theory (perhaps it is more realistic to call it nano-muscle theory, but it is generally known as cross-bridge theory) is that it is possible to study the attachment of heavy meromyosin with isolated actin filaments (Mabuchi, *1991*), so-called decorated actin. Embarrassingly, binding of ATP greatly weakens the affinity of myosin for actin and results

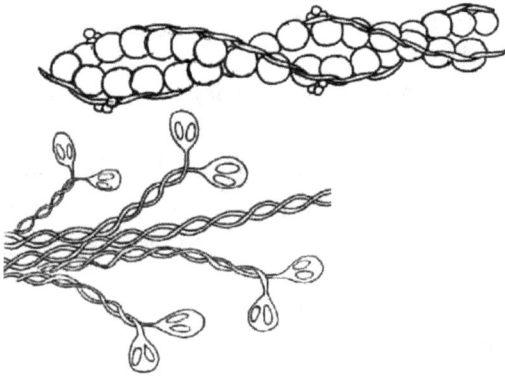

Figure 32: Myosin filaments are polymers helically arranged (top) with heads sticking out (below). Licenced under Creative Commons Attribution-Share Alike 3.0.

Figure 33: Cartoon devised by the author to illustrate the non-electromagnetic theory. The oars represent "cross bridges" which are heavy meromyosin heads 43 nm apart on the thick filament trying to attach to active sites (slits) on the thin filament and pull it along!

in dissociation of the heavy meromyosin attachment. The patterns of decorated actin are described as similar to rigor decoration patterns occurring with scallop proteins. I propose that after death, the time interval before the onset of rigor mortis is a time when ATP and creatine phosphate are slowly depleting, and that that is when the heavy meromyosin attaches to the actin filaments to produce a distorted stiff structure, i.e., rigor mortis. Not something we would like during live muscle activation!

One of the reasons for the popularity of nano-muscle theory is that a favourite preparation for study is the intact single muscle fibre dissected from the frog. This preparation can be tetanised by appropriate electrical stimulation to give long steady traces of force generation. If one adjusts the sarcomere length (SL) at rest to more than SL 2.2 μm, records the force, then during rest, stretch the fibre to a longer sarcomere length, e.g., 2.4 μm, one records a lower force. Continuing this protocol over a range of sarcomere lengths, one obtains a straight-line descending relationship between tetanic force and sarcomere length.

This has been interpreted as meaning that, as one decreases the overlap between the thick and thin filaments, one is also decreasing the number of nano-muscles in apposition and that this is the reason for the decrease in force.

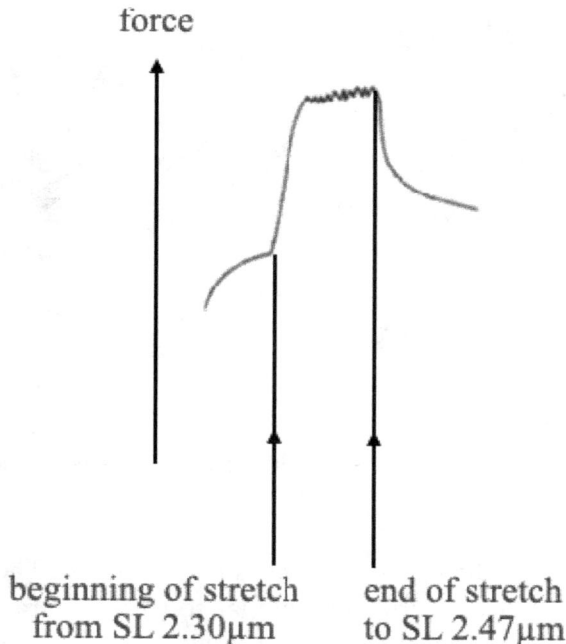

Figure 34:   Tetanised isometric frog skeletal muscle single fibre stretched during activity showing increase in force. The small amplitude force changes during stretch may indicate "jumping" from one peak quantum field to the next. Personal experiment of author. Not previously published.

Unfortunately, if there is ever any difference in the number of heavy meromyosins between the two half sarcomeres, one half sarcomere with heavier meromyosins will pull all the thick filaments right over to the Z disk. This is longitudinal instability to add to the lateral instability already described.

Supposing one stretches the muscle while it is actively developing force? One observes that in spite of decreasing filament overlap, the force increases (Figure 34). As sarcomere length at rest is increased, the increase in force with active stretch becomes greater in spite of the reduction in the numbers of heavy meromyosin heads (nanomuscles or cross bridges) in apposition.

It was argued by some that the procedure caused some changes in the fibre that would lead to a passive interpretation, i.e., that the increased force with active stretch was not a true increase in contractility. The complete description of striated muscle contractility is the force-velocity curve (Daniels, *1984*). Force-velocity curves were therefore measured with active stretch and compared with controls.

These comparisons (Figure 35) confirm an enhanced contractility in spite of less overlap of thick and thin filaments and fewer heavy

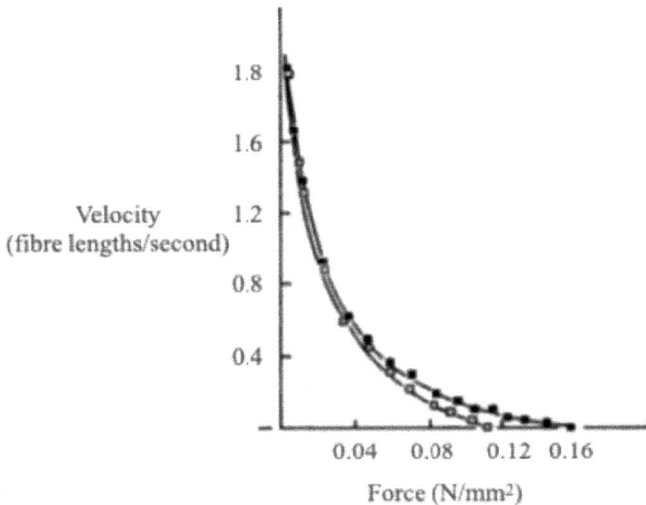

Figure 35: Velocity of shortening obtained from the displacement record during shortening at constant load. Sarcomere length 2.61. Open squares = isometric. Closed squares = after active stretch. Personal experiment of author. Not previously published.

meromyosin heads in apposition to the thin filaments. At each level of force there is enhanced velocity of shortening following active stretch compared with no stretch at the same sarcomere length.

Fact: **All vertebrate muscles increase their force development when stretched during contraction.**

Importance: **This property confers longitudinal stability upon the muscle.**

How can one explain the mechanism of this property?

CHAPTER 17

Different theories

Towards an alternative theory

It should be apparent from Figure 23 that when Ca^{2+} is released from the ends of the SR, the cytoplasmic concentration of Ca^{2+} is higher at the ends of the sarcomere than in the middle (Figure 36). Similar distributions have been found by a number of authors, of which the most recent may be Holash & Macintosh (*2018*).

Imagine actin filaments protruding into the sarcomere from the Z disks encountering this variable $[Ca^{2+}]$ and then being pulled further out. They will be moving into a region of higher $[Ca^{2+}]$. Then suppose that the force exerted on the actin filaments depends on the **local** $[Ca^{2+}]$. As the **local** $[Ca^{2+}]$ is higher, so will the force be higher, i.e., active stretch causes increased contractility (Figure 35).

A solenoid, consisting of a coil through which an electric current is passed, creates a magnetic field into which a rod of suitable composition, e.g., iron, is pulled or exerts a force. If myosin could exert a similar magnetic field and if actin were a magnetisable material, one would have a system suitable for shortening and force development, like a muscle.

The great thing about this method of motion is that the rod **is pulled in a straight line** (Figure 37). There is no sideways motion as one gets with the nano-muscles (cross bridge) theory. **There is no lateral instability.**

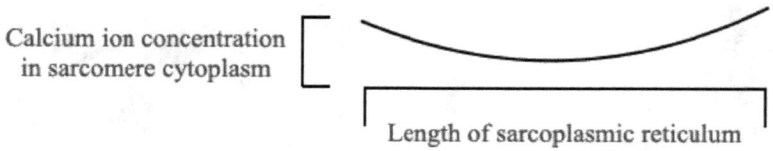

Figure 36:   Simple representation of the variation in cytoplasmic Ca^{2+} concentration during activation along the length of sarcomeres. Not previously published.

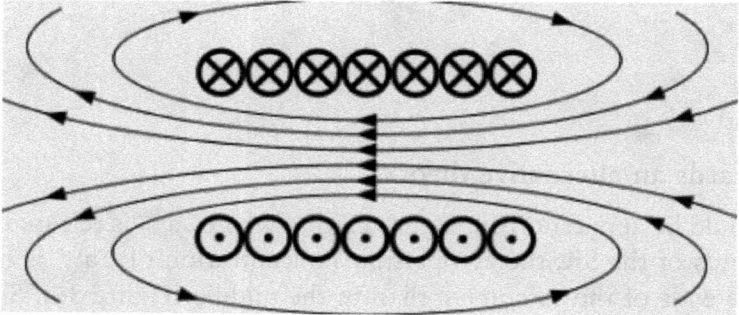

Figure 37:   Electromagnetism of a solenoid coil. Source: Wikimedia Commons, public domain.

CHAPTER 18

Electromagnetic theory of muscle contraction

Lan (*2008*) explored the mechanical properties of individual actin stress fibres in living endothelial cells by nano-indentation using an atomic force microscope. The results demonstrate the pivotal role of such actomyosin contractile levels on mechanical properties. Even more promising, Chen (*2011*) reported magnetic manipulation of actin orientation, polymerisation, and gliding on myosin using super-paramagnetic iron oxide particles. This group showed, for the first time, magnetic manipulation of magnetisable actin filaments at the molecular level, while gliding on a bed of myosin molecules and during polymerisation. But does the heavy meromyosin network provide a suitable magnetic field analagous to that of a solenoid? Possibly, in view of the late Iwazumi's paper (Iwazumi, *1989a*), but he was unable to work out a theory using magnetism, but could do so considering the other aspect of electromagnetism, namely electrostatics. Electrons again.

The snag with the upper scheme in Figure 38 is that if the dielectric constant is less than that of the medium, the rod develops the same polarity surface charges and repulsive force results. This problem does not arise with the magnetic model at all, and in the electrostatic case, it does not arise if the medium is cytoplasmic gel. Iwazumi naturally chose the latter possibility.

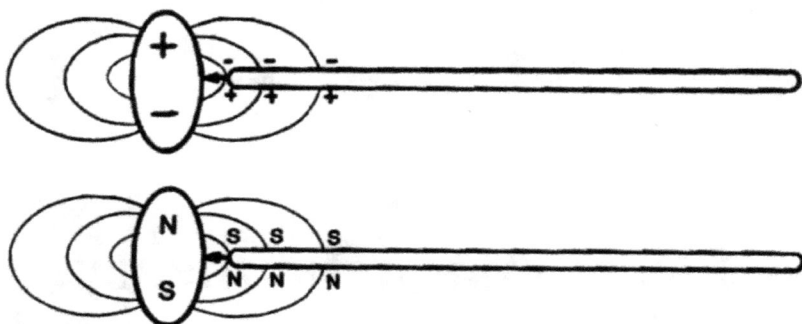

Figure 38:   Above: Interaction between an electric dipole and a ferroelectric rod, resulting in surface charges of opposite polarity on the rod producing an attractive force. Below: Interaction between a magnetic dipole and a ferroelectric rod so that magnetism appears on the rod with opposite polarity producing an attractive force. Reproduced with permission from Iwazumi (deceased) before his death.

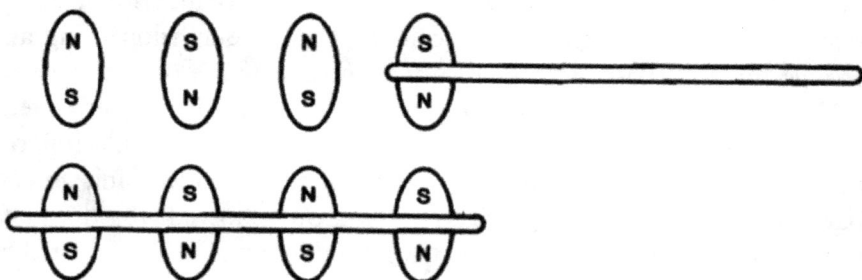

Figure 39:   Above: If a precaution to reduce a friction is not taken, the rod will stick to the magnet and be stuck. Below: When a thin glass plate is inserted between the rod and the magnet, the rod will move to the position shown. Drawing by Iwazumi. Reproduced with permission from Iwazumi (deceased) before his death.

The application of electromagnetic field theory is that the axial force $F = A \times W_{(tip)}$, where A is the effective cross-sectional area of the rod and $W_{(tip)}$ is the energy density (a tensor defined by $E \cdot D/2$ or $H \cdot B/2$ where E, D, H, and B are field vectors commonly used in electromagnetic theory).

Taking the electrostatic equivalent of the situation depicted in the lower diagram of Figure 39 as that of interest in understanding a possible mechanism of contraction, the energy tensor acts on both side surfaces of the rod except at the tip, illustrated by Figure 40.

Figure 40: The field stresses on the side surface do not contribute to the axial force but the sum of all vectors gives an axial force towards the left since no stress appears on the fixed end of the rod. Reproduced with permission from Iwazumi (deceased) before his death.

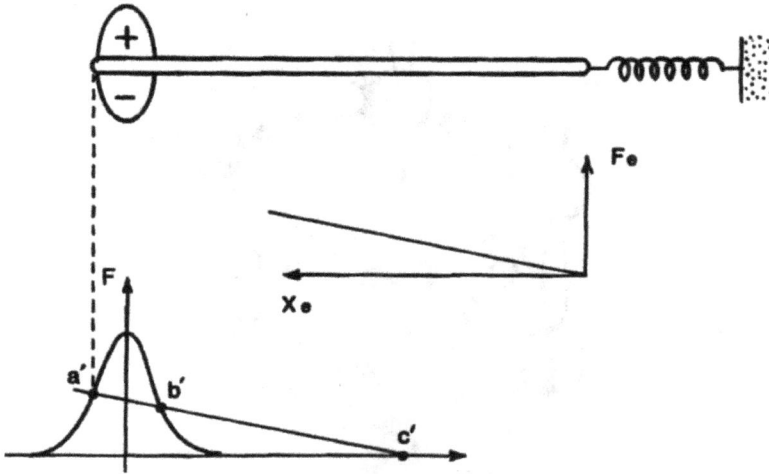

Figure 41: When there is a balance between the field force (Gaussian-shaped curve at the bottom), with the right end of the rod fixed to a wall (maybe equivalent to a Z disk), the spring attachment has a force versus stretch distance (in the inset with stiffness given by the slope — the same slope as line a'-c'). If F is the maximum field force at this location, the force equilibrium points are obtained as intersections of the field force curve and the elasticity line, i.e., a' and b'. The net force is to the left. Reproduced with permission from Iwazumi (deceased) before his death.

I would be surprised if the reader does not wonder whether the rod could be an actin filament and the ± dipole is equivalent to a heavy meromyosin head. An even more realistic representation appears in Figure 41.

In order to emphasise the basic principle of net force in an axial direction, Figure 42 is presented.

The reader should be feeling uncomfortable at this point as this discussion is illustrated in terms of only two dimensions, whereas

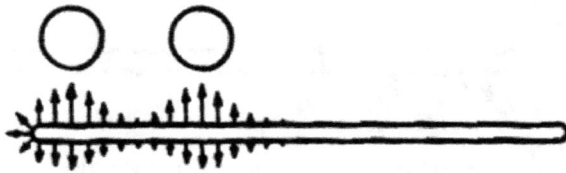

Figure 42:   The side force acting on the rod, net force to the left at the tip, no net force further back on the rod. Reproduced with permission from Iwazumi (deceased) before his death.

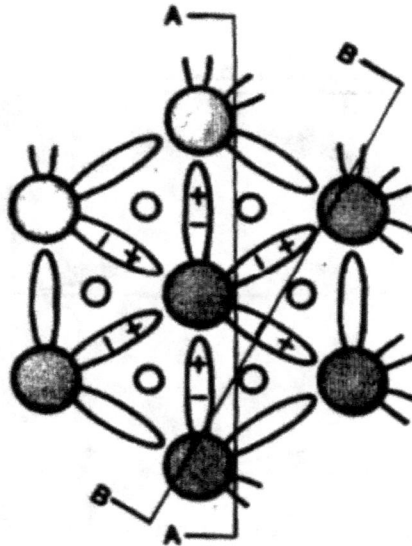

Figure 43:   Transverse arrangement of dipoles and rods. The sections A-A and B-B will be illustrated in Figure 44. Reproduced with permission from Iwazumi (deceased) before his death.

when I was discussing the Vernier effect, I emphasised the need to think in three-dimensional terms.

In transverse section (Figure 43) we see an array of dipoles which produce helical electrostatic fields which have properties to suspend thin filaments in space. What happens if there are different numbers of thin filaments?

Iwazumi (*1970*) studied cross-sectional arrangements of different muscles. These showed two thin filaments in mammalian striated muscle with single overlap. Three thin filaments in the most stable positions were found in insect flight muscle, and the pattern of insect leg muscle was of 6 thin filaments.

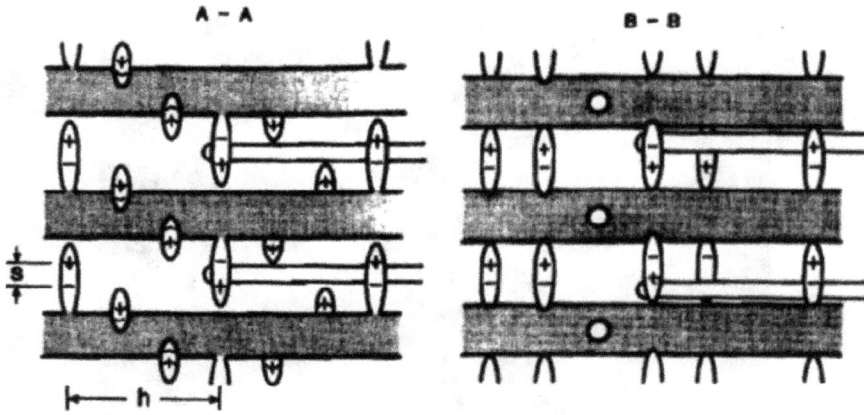

Figure 44:   3D impression of contractile filament arrays from slices A-A and B-B of Figure 43. Reproduced with permission from Iwazumi (deceased) before his death.

Implication: Greater muscle strength is obtained with more thin filaments rather than with more nano-muscles.

Electron micrographic examination of myosin structure of insect flight muscles show the same arrangement as that seen in Figure 43. An alternative representation of the 3-D arrangement is given by the model in Figure 45.

As my main interest is cardiovascular physiology, I now point out that cardiac ventricular muscle has a working range of sarcomere lengths of 1.65 μm to 2.15 μm. If one follows the nano-muscle idea, that would mean that at sarcomere lengths below 2.0 μm, the two 1.0 μm actin filaments will crash into one another — and then what would happen? Would they buckle? In Iwazumi's scheme, they would simply slide past each other into the stable positions in the electrostatic field for two thin filaments. The presence of the shaft of the second filament near the tip of the first in double overlap reduces the intensity of the electrostatic field and therefore less force generation between sarcomere lengths 2.0–1.65 μm, which, together with the increasing Ca^{2+} concentration towards the ends of the sarcomeres, gives heart muscle an inherent ascending limb of the force-sarcomere length relationship (Figure 46; ter Keurs & Noble, 1988). If it had a descending limb as wrongly attributed to skeletal muscle, the heart would rapidly dilate and stop.

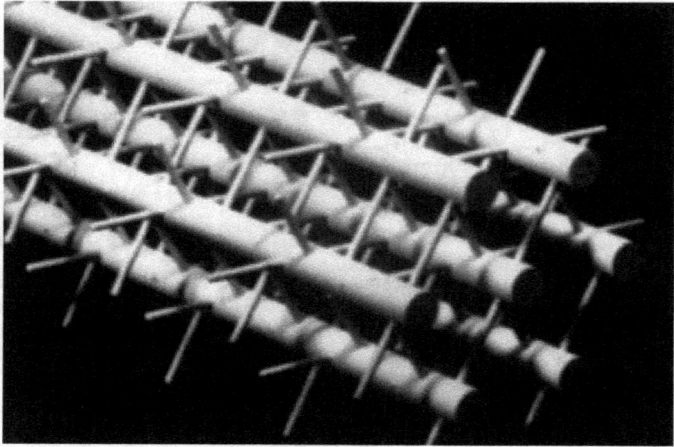

Figure 45:   Model demonstration of the helically rotating electrostatic field, created by the dipoles projecting from the thick filament array, creating the axial stress lines that allow the thin filaments to be suspended in space within the myosin matrix. Reproduced with permission from Iwazumi (deceased) before his death.

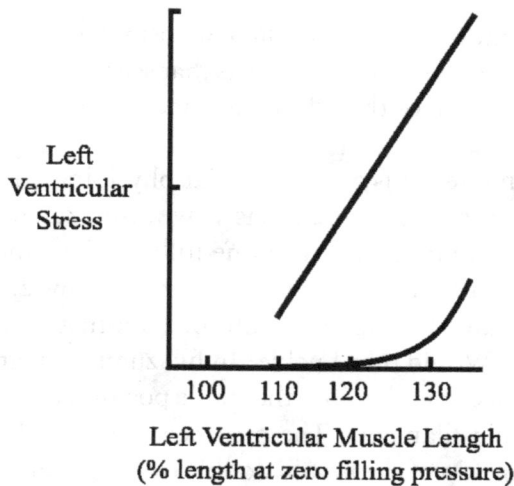

Left Ventricular Stress

100   110   120   130

Left Ventricular Muscle Length
(% length at zero filling pressure)

Figure 46:   Schematic illustration of how the ascending limb of the force/sarcomere length relationship of cardiac muscle (which has no descending limb) translates to the isovolmic pressure/volume relationship (upper line). The lower line is the relationship in diastole (between contractions). From author's teaching notes.

The thin filaments are attached to the cytoskeleton (Figure 41 — the wall), i.e., the Z disks and the contractile system in Figure 41 are envisaged as involving the attachment by a spring. However, we found little evidence of a "series elastic element" in series with the whole muscle in heart muscle (Noble & Else, 1972). The stiffness of muscle increases markedly during contraction (Templeton, 1970, 1972, Templeton & Nardizzi, 1974). However, the observed elastic behaviour is compatible with the stiffness of the connection between thin filaments and the Z disks.

There are many different descriptions of the structure and function of the Z discs which are made of actin. Iwazumi favoured the version of Yamaguchi (1985) who depicted the thin filaments attached at every alternate knot in the Z disc, thus ensuring the required compliance presented to the thin filament. This provides the spring function depicted in Figure 41. For the array of dipoles in Figures 43 and 44, it is required that heavy meromyosin molecules, that are at these positions in muscle thick filaments, should have dipole properties, which is indeed the case (Kobayasi & Totsuka, 1975). Electric birefringence measurements have been made on aqueous solutions of myosin sub-fragments, heavy meromyosin, and sub-fragments 1 and 2 (S-1 and S-2). All of these showed positive electric birefringence. Heavy meromyosin and S-2 showed a large intrinsic Kerr constant, a factor when a strong **electric** field is applied to a fluid. Unfortunately this was not done by Kobayasi & Totsuka in a gel, which would be appropriate for such a study. From the analysis of the build-up and decay process of the birefringence, the contribution of the slow induced dipole moment was concluded in heavy meromyosin and S-2, although the existence of the permanent dipole moment was not completely excluded. The decay process of the birefringence of heavy meromyosin was found to consist of two components, of which the fast one had a relaxation time of the same order as that of S-1. It seems likely that these results could be due to a conformational change in the activated heavy meromyosin head, which has also been detected using X-ray diffraction.

When an electromagnetic radiation is passed through a grid, the received pattern recorded beyond the grid allows calculation of the field spacing. For a grid spacing in striated muscle, the frequencies in the light spectrum are suitable, so that laser diffraction was used; then it became possible to measure sarcomere length in living, contracting and relaxing muscle, e.g., in skinned muscle (Kentish, *1986*) and intact muscle (Daniels, *1984*). For much shorter grid lengths, e.g., the distance of 43 nm between myosin heads (Figure 33), the higher frequency of the X-ray spectrum is used. This method has also been applied to the scattering induced by activation with Ca^{2+} in heavy meromyosin.

Gillian *et al.* (*2013*) found that the binding of Ca^{2+} changes the conformation of the heavy meromyosin molecule from a compact shape found in the "off" state of the muscle to extended relationships between the tail and independently mobile heads that predominate in the "on" state. Their observations are consistent with a Ca^{2+}-free squid heavy meromyosin that is compact, but which becomes extended when Ca^{2+} is bound. Further, the scattering profile derived from the current model of tarantula heavy meromyosin in the "off" state is in excellent agreement with the measured "off" state scattering profile for squid heavy meromyosin. Conformations of regulated myosins or heavy meromyosins from chicken/turkey, scallop, tarantula, limulus, and scorpion sources have been studied by a number of techniques, including electron microscopy, sedimentation, and pulsed electron paramagnetic resonance. These authors concluded that previous studies, and their own studies, together provide "significant evidence that the regulated myosin's compact off-state conformation is an ancient trait, inherited from a common ancestor during divergent evolution".

In my opinion, I would have preferred these studies to have been made in gels, simulating cytoplasm, rather than in fluid simulating the non-existent cytosol. I suppose this is the price we pay for the enormous progress in super technologies, so that we have X-ray diffraction people and nuclear magnetic resonance people and maybe one group is not aware of the importance of both to the generalist trying to use all the available data to try and understand whole physiology.

Iwazumi (*1970, 1989a,* Iwazumi & Noble, *1989*) also had this difficulty. In Iwazumi's 1989 paper, he quotes Morel & Garrigos (*1982*) as showing that the S-1 subunit tends to form dimers even without ATP. The implication of this might be that the myosin network is able to sustain lateral stability in resting muscle with minimal ATP concentration. The trouble is that most studies use aqueous solution rather than gel media; freely suspended molecules in aqueous solution aggregate immediately when they generate strong dipole moments, as we postulate to occur with activation. The focus is on the effect of Ca^{2+} and ATP on the S-1 subunit to cause a change in the orientation of heavy meromyosin with greater separation of electric charge and higher field strength (we do not want electrolysis).

Iwazumi may have got round this problem in theory because, in the vicinity of dipoles, the water (I would say the water in the electrically charged gel) becomes polarised so that it excludes ions.

Figure 47 depicts the hypothesis that the water molecules in the vicinity of the dipole begin to form an ice-like change of structure to reduce potential energy. This structural change reduces entropy of the water, thus causing ions to be excluded and entropic heat to be produced from the polarised region. This idea is compatible with Saenger's (1987) doubts about the behaviour of intracellular water.

The reader may now wonder why there is tropomyosin helically woven into the thin filaments with troponin complexes at 38 nm intervals not aligned with the thick filament myosin heads (Figure 33).

My experience with troponins (troponins C, I and T = the troponin complex) arose in an effort to improve diagnosis and prognosis in patients with acute coronary syndromes. Troponins are protein

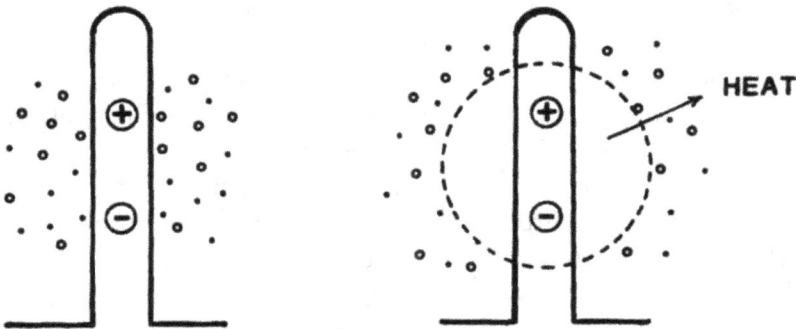

Figure 47: Ion exclusion and entropic heat from water in a strong dipole field. When a dipole appears on the projection, ions are first attracted to either sign of charges, here using conventional polarity titles (electrons negative). Reproduced with permission from Iwazumi (deceased) before his death.

molecules that leak out of damaged cardiac cells. In 1996, a troponin T assay became available with which we were able to make
more accurate diagnoses and a better idea of infarct size that affected
prognosis (Stubbs, 1996a, b). Later, troponin I became the preferred,
more sensitive choice for diagnosis (Keller, 2009) and Hallen (2012)
summarised the supporting evidence in favour of using troponins to
estimate infarct size.

With regard to troponin function in intact healthy muscle cells,
activation of troponin I requires calcium binding to the troponin complex, reversing inhibition of interaction between myosin and actin
in the presence of tropomyosin (Morimoto, 2013). The region of
residues 105–114 (minimum inhibitory peptide), called the 'inhibitory
region', is essential for the inhibitory activity, and the region of
residues 140–148 (second actin-binding region) is also necessary for
full inhibition (Morimoto, 2013). Changes occur in troponin I that
release the inhibition of troponin T and the N-terminal extension
interacts with the N-domain of troponin C but, when phosphorylated
by protein kinase, dissociates from troponin C, decreasing the Ca^{2+}
affinity of troponin C as well as the Ca^{2+} sensitivity of contraction.

What does all this mean in the context of an electromagnetic theory
of contraction? My suggestion is that if the diode array, when activated, exerts force only on the tips of the thin filaments (Figure 40)
and no force on sites away from the tip, then activity and ATP consumption back from the tip are unwanted. This is a similar problem
to the use of Ca^{2+} as activator resulting in unwanted Na^+ and K^+, in
the former case requiring ATP consumption to get rid of Na^+ by the
sodium pump (Figure 16). I suggest that the troponin/tropomyosin
inhibits ATP consumption at sites back from the thin filament tip by
the mechanism outlined in the previous paragraph. This would result
in saving of ATP energy and an increase in efficiency. This mechanism
is predicted to act regardless of the number of thin filaments (double
overlapping in heart ventricle at sarcomere lengths below $2\,\mu$m), as
helical electrostatic fields have properties to suspend thin filaments in
space, with variable position in the cross-section depending on the
number of filaments, as already discussed above.

My attempts in the 1960's to work out, using cross-bridge theory,
how the system worked in cardiac ventricular thin filament double

overlap below $2\,\mu$m sarcomere length was a failure. I failed to get round the result that the cross bridges attaching to the thin filaments pulled them the wrong way, as if trying to pull the overlapped filaments back to single overlap! In the Iwazumi scheme, the filament force and movements are unidirectional, always towards the sarcomere shortening direction with no sideways forces.

In 1989, almost 20 years after my first encounter with Iwazumi, he and I tried to summarise our thoughts and predictions (Iwazumi & Noble, 1989):

1. During contraction, there should be no mechanical motion of the cross projections, which remain helically distributed around the thick filament (Yanagida, 1985).
2. No mechanical contact should exist between the cross projections and the thin filament except during rigor mortis, or the absence of adenosine triphosphate.
3. Sliding between thick and thin filaments is the cause of shortening of sarcomeres (Huxley & Hansen, 1954; Huxley & Niedergerke, 1954).
4. Thin filaments occupy stable positions produced by electrostatic support, positions which are predictable from the number of filaments in the cell (Huxley, 1965).
5. Thin filaments move smoothly and then "jump" 14.3 nm to the next field.
6. The "jumping" of thin filaments would not be possible if the Z-disc was rigid, which would halve the maximum force and reduce the speed of shortening.
7. The descending limb of the length-tension curve in skeletal muscle is due to reduced calcium ions reaching the centre of the A band in the activated sarcomere. Increased sarcomere length in the activated case increases force due to the concave distribution of calcium ions within the A band.
8. Electrostatic induction produces a force simulating non-linear viscous drag on the thin filament (Abbott & Aubert, 1952).
9. Both quick-release and quick-stretch responses will consist of three components (Ford, 1977).

10. Velocity of shortening depends on the speed of field development of the dipoles ahead of the tips of the thin filaments. Since the level of force and the time to reach the level are proportional, the relationship between force and velocity is inverse and hyperbolic except near the isometric state with tips in the central bare zone.

11. Rigor binding of myosin and the S1 sub-unit fragments to actin have no relation to the process of normal contraction.

12. The myosin head produces a dipole field when reacted with adenosine triphosphate (ATP). Splitting of ATP results from positive escape of surface charges. The dipole will be maintained if there is no escape of charge.

13. The conversion ratio of ATP to mechanical energy will not be fixed.

14. Absence of magnesium results in loss of force in spite of increased ATP, because of loss of apparent dielectric constant of thin filaments.

15. Actin monomer with a low dielectric constant will not stimulate myosin ATPase, whereas actin polymer (filament) will do so (Offer, 1972).

16. The initial burst of consumption of ATP is used to polarise the water around the myosin head and to drive out the ions.

17. The lag time between the rise of intracellular Ca^{2+} and the rise in tension (Figure 19) is due to the time required to polarise the water around the myosin heads.

18. Reduced ionic strength will increase force and maximum velocity, and *vice versa*.

19. Excess heat at the beginning of contraction and heat absorption afterwards (Curtin & Woledge, 1978) is due to changes in entropy in water.

20. Mass transfer detected by X-ray diffraction will be dependent on the presence of potassium ions.

Since the sad demise of Iwazumi after all this was stated, I would like to add:

21. The actomyosin/troponin complexes of the thin filaments act to reduce ATP splitting where it is not needed back from the tip of the thin filament.

22. The release of the Ca^{2+} activator depends upon the opening of a cell membrane channel by an outflow from the cell to the extracellular compartment of electrons and release of stored Ca^{2+} from the SR. The lost electrons are replaced by electron flow from mitochondria to the cytoplasm.

23. In our joint paper (Iwazumi & Noble, *1989*), we called the title an electrostatic theory, unaware that Einstein had shown that electrostatic fields and electromagnetic fields are not different (as Maxwell thought), but are described by the same mathematical equations (Rovelli, *2014*). We now just have electromagnetic fields and now postulate that muscle contraction is achieved by a force on thin filaments exerted by electromagnetic quantum fields.

| CHAPTER 19 |

Electrophysiology of smooth muscle

Vascular smooth muscle

Smooth muscles lack the kind of "all-or-nothing" responses of cardiac and skeletal muscles. Cardiac muscles have to contract and relax continuously for a lifetime and do so even in the absence of its sympathetic and parasympathetic innervation (Noble, *1972*). Skeletal muscles have to contract rapidly to nerve impulses from the motor neurones and relax when they slow or stop. Both these circumstances require very ordered arrays of thick and thin filaments accounting for their striped (striated) appearance. Smooth muscles have to contract continuously ("tone"), although with changes in force generation, and have a less ordered arrangement of the contractile filaments of myosin and actin (non-striated versus striated of cardiac and skeletal muscles). They occur in the walls of various tubular structures and control the lumen of these structures; for instance, the hydrodynamic and aerodynamic resistance in bronchi.

The autonomic nerve endings are very important in that the release of the neurotransmitters, noradrenaline and acetylcholine, exert control of function in many organs including blood vessels, i.e., arteries, arterioles and veins. In arteries the smooth muscle in the arterial wall is constantly being stretched and released with the changes in intraluminal pressure; you can feel the pulsation of an artery. As was the case with striated muscle (Figure 34), smooth muscle with tone, when stretched, increases its force of contraction (more activator Ca^{2+} at the ends of the sarcoplasmic reticulum compared with the centre, Figure 37). Therefore, one postulates that the efficiency of arterial smooth muscle will be increased by the fact that it is stretched during each systole (Bayliss, 1902). The effect of this is called myogenic autoregulation, which increases with muscle length and is greater at higher rates of stretch (Johansson & Mellander, 1975). This idea of an extra contribution, demonstrated by Bayliss in the the portal vein and by Mellander & Arvidsson (1974) in the resistance vessels that determine arterial pressure, was studied more recently by Markos (2012) in the porcine iliac artery, i.e., a conduit artery.

One way of assessing the effect of pulsatility on tone would be to block out the pulsatility in a segment of artery and see whether the cross-sectional area of the segment changed, and if so, in which direction and by how much. This was considered to be extremely

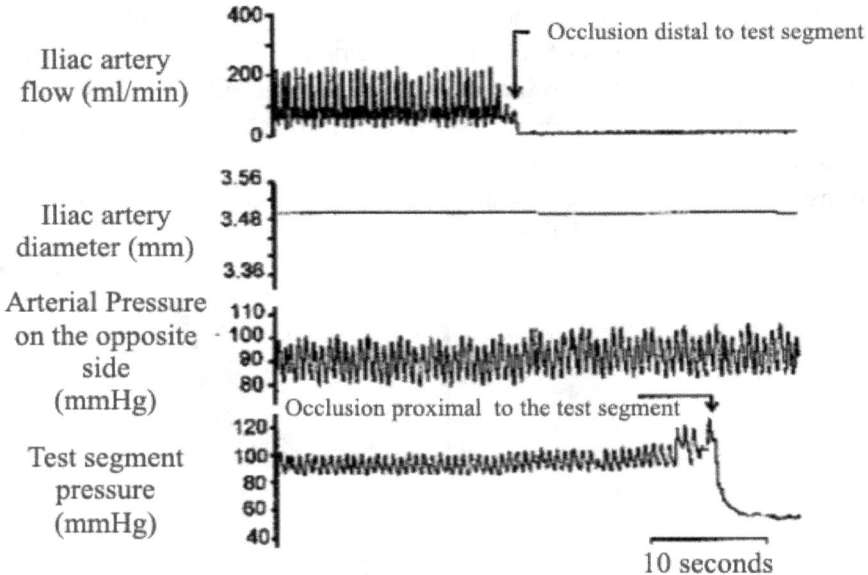

Figure 48:   Recording from an anaesthetised pig with constant systemic arterial pressure (3rd trace). A test segment of iliac artery was occluded, first distally, then proximally, at constant diameter. At the same diameter, there was a marked fall in the pressure within the segment when pulsatility was removed by the proximal occlusion, indicating lessening of contractile tone. From an experiment devised and performed by the author with assistance. Not previously published.

difficult to perform in an intact anaesthetised animal, as it would require complicated artificial "plumbing" of the artery and perfusion with non-pulsatile pumps, anticoagulation, etc. With colleagues in Markos's group, we decided to simply isolate the segment, keeping the diameter constant, and observe what happened to the pressure in the segment (Figure 48).

We thought that they had demonstrated an effect of loss of pulsatility upon the pressure at constant diameter, but went on to examine what happened when they injected blood into the segment with distal occlusion (pulsatility still present) and with distal and proximal occlusion (no pulsatility). Their overall results suggested that there was an effect on the pressure/diameter relationship consistent with loss of tone when pulsatility was removed.

It seems a reasonable proposal, in view of data in Figure 49, that tone is partly maintained by muscle stretch enhancement of force during pulsatility (Markos, 2012).

In Figure 50, it is assumed that the small changes in incremental compliance (the slopes of the lines in the figure) can be neglected and that the diameter-pressure line is straight (it is actually convex to the diameter axis). Smooth muscle relaxation shifts the diameter-pressure relationship, from the control position to the left, when achieved at constant diameter by removing pulsatility, and upwards at constant pressure with increased shear stress (Kelly & Snow, 2011).

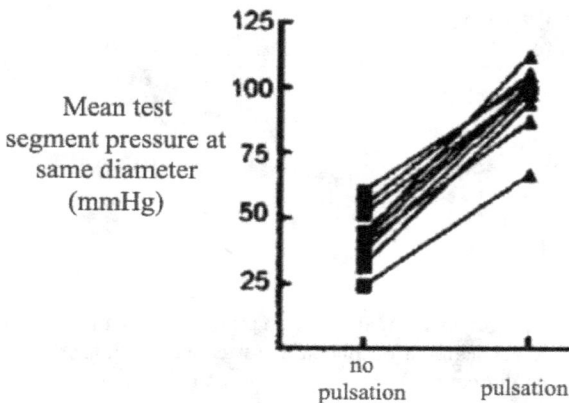

Figure 49: Demonstration in several experiments of the difference in pressure at constant diameter of the presence or absence of pulsatility. Not previously published.

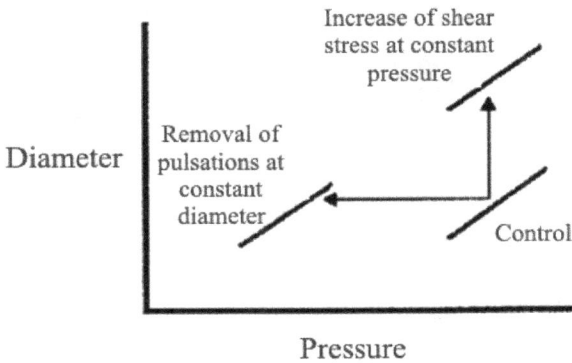

Figure 50: Simplified model of the effect of arterial smooth muscle relaxation in the diameter-pressure domain.

It is important to realise that the electrophysiology of smooth muscle to be described consists of events that take place on top of this myogenic autoregulation.

The blood vessels dilate in response to demands for increased blood supply in the activated organ concerned. A common such situation is exercise in which the skeletal muscle requires an increased supply of oxygen and substrate from the blood. However, exercise is associated with activation of the sympathetic nerves which release noradrenaline at their nerve endings, and that depolarises the smooth muscle (below the resting electron density of 60 mV — similar to the sinus node cell), causing an increased Ca^{2+} intracellular concentration due to the Ca^{2+} current and SR release; this is vasoconstricting (by activating actomyosin) and reduces blood flow. Indeed, under resting conditions, there is some of this vasoconstriction, known as "tone", with the electron density at about 60 mV (from which electrons can outflow slowly as in sinus node cells).

CHAPTER 20

Flow-mediated dilatation

Role of nitric oxide in restoring electron density

This "unwanted" effect (reflex adrenergic vasoconstriction) during increased organ demand is overcome by a different system. Variation of electron density and thus of tone is mainly achieved under these circumstances through nitric oxide (NO) production by the NO synthase of the endothelium (endothelium-dependent relaxing factor, EDRF). A very important aspect of vascular smooth muscle is the maintenance of vessel tone, and, in the case of arteries, vasodilation to accommodate increased flow in response to increased downstream demand, so-called flow-mediated dilatation (FMD) (Markos, *2013*).

Possible mechanisms for FMD (Figure 51) are: (i) conducted vasodilation from the periphery mediated by endothelial cell hyperpolarising factor; (ii) relaxation of smooth muscle tone by nitric oxide produced by the endothelial cells stimulated from distortion of endothelial cell cytoskeleton; and (iii) as for (ii) with the cytoskeleton deformation transmitted from the glycocalyx, the gel layer between the arterial endothelium and the blood.

Hilton (*1959*) excluded a neural reflex mechanism and it was thought that an electrical signal was conducted through the smooth muscle layer because the conduction speed was slow, of the order of 10 cm/s. The theory was reasonable, due to the known existence of endothelium-derived hyperpolarising factor (EDHF, still not defined but I suggest that it is electron flow through tissue — see below),

Circumferential stretch

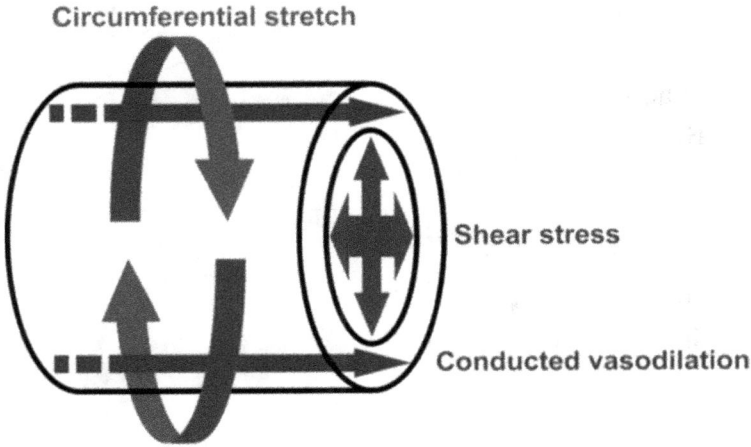

Shear stress

Conducted vasodilation

Figure 51:   Schematic diagram of a conduit artery. An increase in blood flow, due to increased demand downstream, causes an increase in wall shear stress (double-headed arrows), which is the stimulus for arterial dilatation (flow-mediated dilatation), accompanied by a reduction in circumferential smooth muscle tone (curved arrows). Flow-mediated dilatation is opposed by basal muscle tone and external pressure.

which spreads through gap junctions between endothelial and smooth muscle cells along the arterial wall. An important factor in the relevance of Hilton's theory as a mechanism for FMD is the question of the distances possible for smooth muscle hyperpolarisation. The assumption is that such distances are finite, otherwise hyperpolarisation at one site of the arterial tree would spread to the whole tree. This has been disproved; however, one concern with these findings is that when measurements are made *in vitro*, excised arteries are bathed in physiological solutions that are highly electrically conductive. One of the most thorough explorations of FMD was conducted *in vivo* by Kelly & Snow (*2011*) in the iliac artery of an anaesthetised pig; the iliac artery is located much more than 3 mm away from the peripheral femoral bed. The results obtained from those studies contradict those of Hilton, because the iliac artery and vein were connected via an adjustable shunt that allowed discrete, stepwise changes in flow, shear stress and arterial diameter. Crucially, FMD still occurred, when the peripheral bed was excluded in the presence of the shunt.

FMD is illustrated in Figure 52. The vasodilation is mediated by nitric oxide (NO) released from endothelial cells, because it is blocked

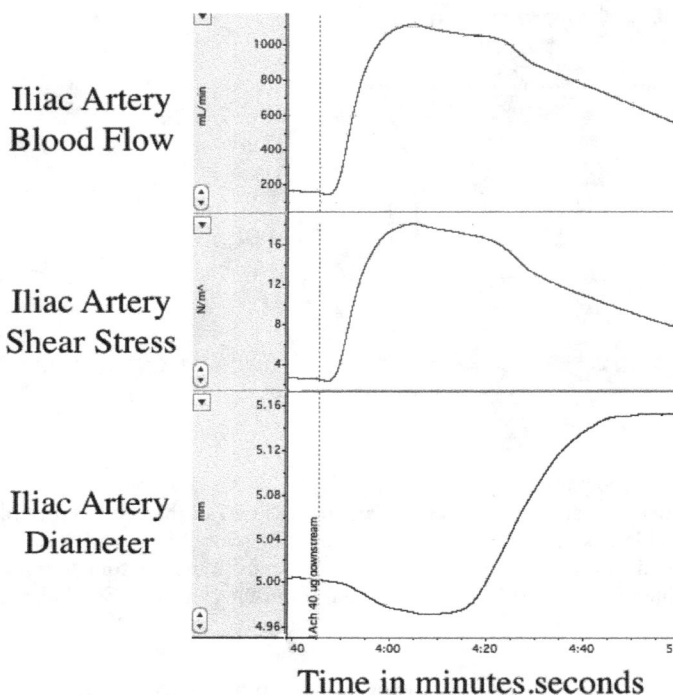

Time in minutes.seconds

Figure 52:   The effect of an increased shear stress on the interior of a conduit arterial wall caused by increased flow due to increased demand for oxygen and substrate of the downstream organ – usually named flow-mediated dilatation. From an experiment performed by the author with assistance. Not previously published.

by NO synthase inhibitors. NO also maintains some control of electron density during control conditions (so-called hyperpolarisation), which is reversed in some species by N-nitro-L-arginine, which blocks NO synthase. The dependence of hyperpolarisation on cellular potassium ions and the opening of calcium-activated potassium channels have led to the hypothesis that the increased electron density occurs through an outward flow of potassium ions. However, this would create an unsteady state of potassium ions and disrupt the ionic equilibrium. Alternatively NO inhibits the calcium ion inward current (Zima, 1996), which is accompanied by reduced electron outflow. At the same time any drop in electron density will cause increased electron flow from mitochondria. The vascular smooth muscle cell is an example of control of function by varying a continuous generation and outflow of electrons with varying cellular "negative" charge, including some spikes. The data of Bulbring (1985) is compatible with this theory. Disproof of the theory can be achieved by measuring the extracellular potassium ion transient in response to NO and determining whether it is large enough to account for all of the accompanying hyperpolarisation.

The vital role of vascular endothelial cells in control of vascular smooth muscle cell depolarisation and repolarisation through the NO system is generally recognised. At normal extracellular K^+ concentration this sustains an electron density close to 80 mV. The glycocalyx layer that separates the vascular endothelial cell from the blood is negatively charged, forming an electrostatic barrier for plasma cells and proteins, like albumin. However, the negative charge of albumin is part of the electrostatic barrier along with negatively charged heperan sulphate. The electrostatic barrier provided by the endothelial cell glycocalyx may also repel adhesion of blood cells. The actual electrical potential of the glycocalyx does not seem to have been measured. In addition, the glycocalyx participates in the functions of the endothelial cell in helping to sense increases in shear stress at the vessel wall induced by increases in blood flow, thus triggering NO production and vasodilation (VSM relaxation). That endothelial cell depolarisation (outward flow of electrons), associated with reduced cell stiffness, is possibly involved in shear stress-induced vasodilatation requires experimental testing. Such a test of this would be to monitor endothelial cell trans-membrane potential (electron density) changes during increased blood flow.

The mechanism by which FMD occurs is still the subject of debate. Three main theories are described below.

- Theory A: A retrograde hyperpolarisation from the arterioles, which ultimately results in dilatation of the smooth muscle of the feeding conduit artery (Hilton, 1959).
- Theory B: The cytoskeleton of the endothelial cell directly senses changes in wall shear stress. This causes mechanical changes in elements of the cytoskeletal structure leading to the production of NO, which relaxes the smooth muscle of the artery (White & Frangos, 2007).
- Theory C: The arterial glycocalyx, a 0.5 μm gel located between the endothelial cells and the blood, senses the changes in wall stress,

which, in turn, causes mechanical changes in elements of cytoskeletal structure leading to production of NO (Weinbaum, *2003*).

My opinion about this tends towards Theory C because:

1. FMD is inhibited by hyaluronidase which degrades hyaluronic acid glycosaminoglycans within the glycocalyx layer.
2. FMD is inhibited by increased luminal dextrose concentration but not other monosaccharide concentrations, e.g., l-glucose and mannitol (Kelly, *2006*), implying decreased glycocalyx function due to increased glycosylation of glycocalyx glycoproteins.

The glycocalyx-glycoproteins are a continuous part of the cytoskeletal structure. Thereby they make the lipid-based cell-membrane part of the cytoskeleton sensing internal and external strain. This would provide a mechanism for electron flow in response to membrane strain. In my opinion, the more conclusive opinion would remain obscure until we know what happens to endothelial cell electron density during FMD. I would expect that electron outflow would occur at some stage, because the resting value is about 80 mV. Perhaps there is electron outflow when shear stress increases, followed by endothelial cell depolarisation, which stimulates NO synthase NO production?

CHAPTER 21

Pulmonary vessels

Role of hypoxia and serotonin

The smooth muscles of pulmonary vessels are different from that of systemic vessels, presumably because of their specialised role at birth of changing from constriction to dilatation when the lungs inflate with air.

Later they fail to respond initially from 50–60 mV resting electron density upon sympathetic stimulation (Figure 53). They depolarise and vasoconstrict strongly in response to hypoxia (opposite of systemic vessels) and serotonin, e.g., released from venous thromboses.

Although vascular smooth muscle is primarily innervated by the sympathetic nervous system, parasympathetic stimulation is also claimed to play a role in smooth muscle cells. The neurotransmitter at parasympathetic nerve endings is acetylcholine. This is a potent vasodilator when injected, by inhibition of electron flow and tone, but its role in normal physiology of vascular smooth muscle is unclear because of the more major effect of NO.

The systemic veins act as a blood reservoir, so that sympathetic activation and consequent smooth muscle contraction initiated by depolarisation (electron outflow) cause blood volume in the heart to increase, and thus increase cardiac output. By contrast the arterioles dilate during exercise under the influence of NO-induced hyperpolarisation, reducing peripheral resistance which also increases cardiac output.

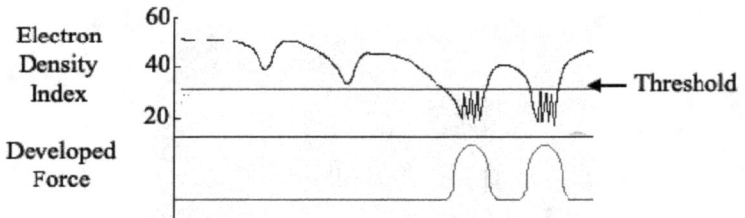

Figure 53: Spontaneous variations in pulmonary smooth muscle electron density (index in mV) occur in spite of some sympathetic stimulation until the threshold is reached for action potentials and contractions. Electron density changes not previously published.

CHAPTER 22

Non-vascular smooth muscle

Gut

The electron density of most smooth muscle cells in the gut (Figure 54) and walls of ducts is between 50 and 60 mV. In contrast to nerves and striated muscle cells, the electron density of smooth muscle cells fluctuates spontaneously. Slow wave activity appears to be a property intrinsic to smooth muscle and not dependent on motor neurone control. Innervation by sympathetic and parasympathetic nerves modify the intrinsic activity.

Normal gastrointestinal motility (peristalsis) results from coordinated contractions of smooth muscle, which in turn derive from two basic patterns of electrical activity across the membranes of smooth muscle cells — slow waves and spike potentials (action potentials). The contractile phase of the slow waves is induced by electron outflow from electron dense cytoplasm. There are also more rapid contractions induced by spikes of electron outflow. The gut smooth muscle cells are electrically coupled, so the fluctuations in electron density spread to adjacent sections of muscle, resulting in the slow waves of partial depolarisation in smooth muscle that sweep along the digestive tube for long distances. These partial depolarisations are equivalent to fluctuations in electron density of -5 to -15 mV from about 60 mV. The frequency of slow waves depends on the section of the digestive tube; in the small intestine, they occur 10–20 times per minute and in the stomach and large intestine 3–8 times per minute. The larger spike electron outflows, when they occur, do so at the crest

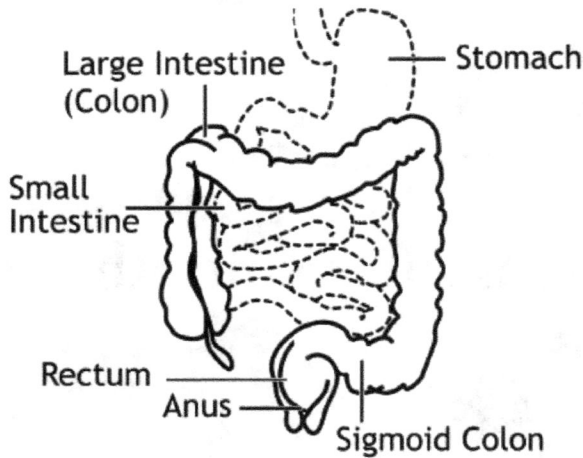

Figure 54: The gut, a series of tubes containing smooth muscles in their walls. Source: Wikipedia, public domain.

of slow wave depolarisations. Spike potentials result when a slow wave passes over an area of smooth muscle that has been primed by exposure to neurotransmitters released in their vicinity by neurones of the sympathetic nerve endings (noradrenaline) causing more electron outflow, and contraction of parasympathetic nerve endings (acetyl-choline) causing more electron inflow from mitochondria. These neurotransmitters are released in response to a variety of local stimuli, including distension of the wall of the digestive tube, and serve to "sensitise" the muscle by making its resting electron density less positive.

To summarise:

1. The bolus distends the gut, stretching its walls.
2. Stretching stimulates nerves in the wall of the gut to release neuro-transmitters into smooth muscle at the site of distension — the electron density of that section of muscle becomes more depolarised (electron loss). There may also be stretch activation of the muscle.
3. Slow waves consist of contractions of circular muscle, in what is called the propulsive segment (no one, to my knowledge, has suggested that this is caused by an inward Na^+ current!). When a slow wave passes over this area of sensitised smooth muscle, spike potentials form and contraction results. The slow wave continues

with a dilation and relaxation of the constriction in the propulsive segment; this is called the receiving segment (no one, to my knowledge, has suggested that this is caused by an outward K^+ current!). It should be evident that continuous involvement of Na^+ and K^+ currents would lead to unstoppable ionic imbalance.

4. The contraction moves around and along the gut in a coordinated manner because the muscle cells are electrically coupled through gap junctions, i.e., by electricity, which equals electron flow by definition. Gap junctions are clusters of hydrophilic intercellular channels formed by the docking of two so-called connexions, each contributed by a neighbouring cell; this allows the direct transfer of signalling molecules and provides a pathway of low resistance for the spread of electrical currents, i.e., electron flows between cells.

Implication: **If coordination of contraction and relaxation in the gut is achieved electrically, why should it not be the same in other muscles, e.g., heart muscle?**

Defecation is a voluntary act that finally gets rid of the food residue after extraction of nutrients. The internal anal sphincter is controlled by parasympathetic fibres which relax the sphincter involuntarily. The external anal sphincter is skeletal muscle that is controlled by somatic nerve supply from the inferior anal branch of the pudendal nerve (S2, 3, 4), which allows conscious control of defecation.

Ducts

The bile duct has the important function of delivering liver products and enzymes into the gut. This seems to be achieved passively (Wakim, 1971) until the secretions arrive at the junction with the gut where there are biliary sphincters (Torsoli, 1971); most studies focus on the sphincter of Oddi (SO). The primary functions of the SO are to regulate the delivery of bile and pancreatic juice into the duodenum, and to prevent the reflux of duodenal contents into the biliary and pancreatic systems (Woods, 2005).

During fasting, most of the hepatic bile is diverted toward the gallbladder by the resistance of the SO. The gallbladder allows the gradual entry of bile, relaxing by passive and active mechanisms. During the digestive phase the gallbladder contracts and the SO relaxes, allowing bile to be released into the duodenum for the digestion and absorption of fats.

Regulating the motility of the gallbladder are the vagus and splanchnic nerves and the hormone cholecystokinin (Behar, 2013). The vagus nerve contains afferent and efferent fibres; the latter are pre-ganglionic neurones that synapse to intramural postganglionic cholinergic neurones present within the gallbladder wall. Stimulation of the efferent fibres of the vagus nerve contracts the gallbladder that is antagonised by the ganglionic blocker hexamethonium, and atropine, the muscarinic receptor antagonist. Stimulation of the splanchnic nerves causes a relaxation of the gallbladder that is blocked by propranolol. This inhibitory innervation was demonstrated in atropine-treated cat gallbladder muscle strips. The strips relaxed in response to electrical field stimulation that was blocked by propranolol but was unaffected by partial antagonists of the vasoactive intestinal peptide, suggesting that the stimulus was acting on sympathetic postganglionic neurones releasing catecholamine as the neurotransmitter acting on beta-adrenergic receptors. However, different results were obtained when the atropine-treated cat gallbladder was relaxed by stimulating the vagus nerve.

All this indicates the great importance, for this circular arrangement of smooth muscle, autonomic nerves, cytokines and gut contents, in affecting the sphincter's contraction and relaxation. However, there seems to be no strong reason to doubt that, in the final action, as with other smooth muscle (e.g., gut), the contractions are achieved through electron outflow to cause reduction of electron density, and relaxation to be achieved by restoration of electron density by electron flow from the mitochondria to the cytoplasm, all coordinated electrically through gap junctions. The electron density of gallbladder, common bile duct and SO vary between 40 and 56 mV, but may decline to lower values during contractions.

There seems to be little information about the electrical aspects of the smooth muscle cells in the pancreatic duct, of which the opening into the gut of pancreatic enzymes also involves the SO. In salivary glands and other exocrine organs, there are starfish-shaped cells that lie between the basal lamina and the acinar and ductal cells (Redman, 1994). These have structural features of both epithelium and smooth muscle cells, and so are called myoepithelial cells. Their functions include contraction when the gland is stimulated to secrete, and compressing or reinforcing the underlying parenchymal cells, thus aiding in the expulsion of saliva and preventing damage to the other cells. They also may aid in the propagation of secretory and other stimuli. No electrical measurements have been made in these, but I would expect that the contractions in these myoepithelial cells are triggered by electron outflow.

Bladder, ureter, urethra and urinary sphincter

Ultrasound studies of bladder filling have shown that tonic 'closure' of the ureters is interrupted ~every 10 minutes by urine expulsion from the renal pelvis into the bladder. Smooth muscles from the urethra and bladder (Figure 55) display characteristic patterns of spontaneous contractile activity in the filling phase of the micturition cycle. Tonic contractions are seen in the urethral smooth muscles, and phasic contractions occur in the detrusor. The common function of the bladder in mammals is to store and expel urine; there should therefore be underlying similarities in the properties of the detrusor in all species. The detrusor muscle, also detrusor urinae muscle, muscularis propria of the urinary bladder and (less precise) muscularis propria, is smooth muscle found in the wall of the bladder. The detrusor muscle remains relaxed to allow the bladder to store urine, and contracts during micturition to release urine (Brading, 2006). The sphincter is striated muscle.

Micturition is normally a voluntary act, achieved by innervation of the lower urinary tract by 3 sets of peripheral nerves: pelvic parasympathetic nerves; nerves which arise at the sacral level of the spinal cord, excite the bladder, and relax the urethra; and lumbar sympathetic nerves, which inhibit the bladder body and excite the bladder base and urethra. When the bladder is full of urine, stretch receptors in the bladder wall trigger the micturition reflex. The detrusor muscle that surrounds the bladder contracts. The internal urethral sphincter relaxes, allowing urine to pass out of the bladder into the urethra. In spite of this simple explanation, it is a complex system of autonomic

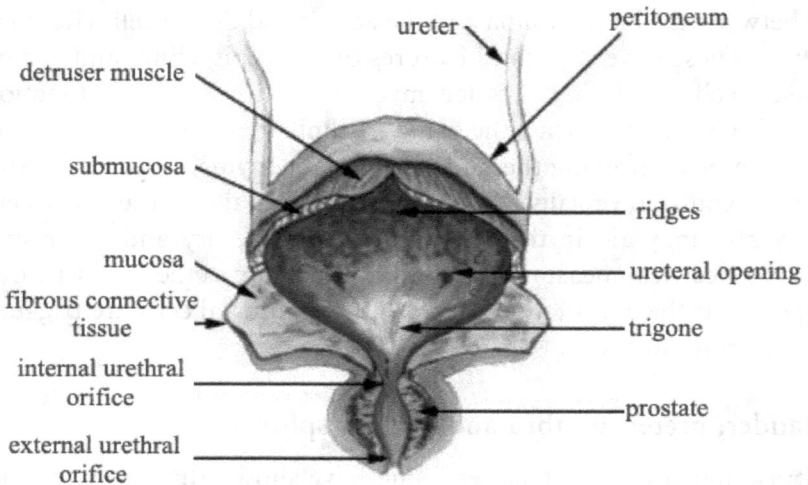

Figure 55: Smooth muscles of the urinary system, of which the major one that effects bladder emptying is the detruser muscle. Source: Wikimedia Commons, public domain.

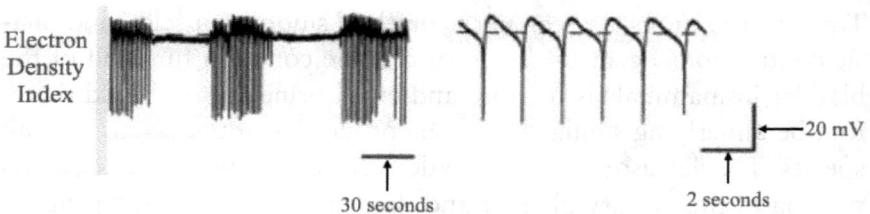

Figure 56: First depiction of electron outflows leading to electron loss action potentials in detruser muscle. Action potentials from human detruser muscle that occur in bursts (left) show slow depolarisation until a threshold is reached at which the action potentials are fired (right). Not previously published.

and somatic nerves, the former of which control the smooth muscle of the bladder and the urethra, the latter of which activate the striated sphincter (Fowler, 2008).

Resting electron densities of the detruser muscle, determined at the most positive, stable level between each action potential, range between 39 and 50 mV.

There is a slow drift downwards of electron density, sometimes resulting in bursts of action potentials (Figure 56). This is similar to the pacemaker current in sino-atrial node cells during diastole. When bursts of action potentials are generated, they occur at a frequency in the range between 0.8 and 2 per minute and each burst consists of 3–20 action potentials (Hashitani & Brading, 2003). As with the sino-atrial node cells, the action potentials are Ca^{2+} dependent.

Intracellular electrical recordings have been made from urethral and anal sphincters but they lack the clarity of those recorded from detruser muscle (Figure 56) and I hesitate to offer an interpretation.

CHAPTER 23

Hierarchy of vertebrate muscle

All muscle types are postulated to have in common:

(1) Generation of electricity by mitochondria
(2) Depolarisation and repolarisation by outflow of electrons from cytoplasm, followed by restoration of cytoplasmic electrons from mitochondrial electron generation
(3) Depolarisation-induced opening of the Ca^{2+} channel and Ca^{2+} inflow
(4) Ca^{2+} extrusion by NCX and possibly in some cases by a sarcolemmal ATPase
(5) Ca^{2+} release and uptake by the SR; uptake by the Ca^{2+}-induced SR ATPase
(6) Electrical conduction and coordination via gap junctions with the possible exception of some skeletal muscles

  (a) *Most active*: cardiac ventricular muscle and Purkinje fibres with most mitochondria, highest resting (diastolic) electron density, highest electron outflow, greatest Ca^{2+} cycle, greatest electron inflow from mitochondria, highest energy consumption (work, heat, electricity)

  (b) *Next most active*: skeletal muscle with somewhat less of all of the above features

(c) *Least active*: sinus node, atria, vascular smooth muscle, gut and duct smooth muscle

Generation ⟶ electron flow ⟶ consumption ⟶ heat

The electrical system is one way: This is the same as with man-made electricity and completely avoids the difficulty of maintaining balance that attends any ionic flow system, e.g., Ca^{2+} balance.

CHAPTER 24

Electrophysiology
of endocrine glands

There is much less information about the electrophysiology of endocrine glands compared with nerves and muscles. The presence of resting intracellular electron density, sensitivity to external chemical environment, and involvement of Ca^{2+} as second messenger, suggest that the fundamental principles usually apply, i.e., depolarisation due to electron outflow, reduction of cell membrane electric field force and consequent opening of the Ca^{2+} channel, Ca^{2+} inflow triggering function and then restoration of ionic balance. The method of experimental investigation of depolarisations frequently involves changes in extracellular K^+, which is not the situation in the intact animal with a relatively constant value of extracellular K^+. The most common secondary messenger appears to be Ca^{2+} that releases the contents of the secretary organelles containing those specialised molecules that effect the specialised functions.

Pituitary-hypothalamus complex: "The conductor of the endocrine orchestra"

Much hormonal control is exerted by the secretion of hormones that stimulate other endocrine glands such as the thyroid and adrenal glands. Action potentials result from spontaneous slow outward electron flows from initial electron densities of 50–60 mV, followed by spikes on reaching a threshold level of electron density. These cause inflow of Ca^{2+} resulting in spike-like action potentials that trigger hormone secretion. There are then rapid or delayed repolarising flows of electrons from mitochondria to cytoplasm (Stojilkovic, 2012). The electrical activity is independent of external stimuli. This is a similar mechanism to that described for sinus node cells above, although not as regular. The description of the depolarisations as spontaneous, rather than Na^+ inflows, argues for them being simple electric currents (i.e., electron flows).

An exception to the general ignorance of electric events in the various types of glands is the islet cells of the pancreas; this will be treated first.

Electrophysiology of the endocrine pancreas

The pancreas has acinar cells with an electron density of approximately 40 mV that secrete digestive substances into the gut, and β islet of Langerhans cells in which the electron density is highly sensitive to blood glucose concentration, and which secrete insulin into the blood, a Ca^{2+}-dependent mechanism (Dean & Matthews, 1970).

Figure 57 shows falling electron density (dashed line) due to electron outflow which reaches a critical range of glucose concentration (shaded area) containing the threshold for action potential discharges (dots and continuous line) over the normal range (SDs at lower and upper limits). At glucose concentrations above this range a large increase in action potentials occurs in the "diabetic" range. Action potential frequency is thought to determine insulin secretion rate. Failure of this control system results in diabetes mellitus.

I postulate that the electron outflow depolarisation (dashed line in the picture) that initiates this system is caused by a reduction in cell membrane impedance caused by increasing glucose concentration, an example of chemically induced depolarisation. The mechanism is Ca^{2+} sensitive, implying that the Ca^{2+} channel opens allowing Ca^{2+} entry, followed by NCX and sodium pump activity. Dean & Matthews (1970) concluded that the action potential induced by glucose in islet β-cells is due predominantly to calcium entry and that sodium ions tend to repress this calcium influx.

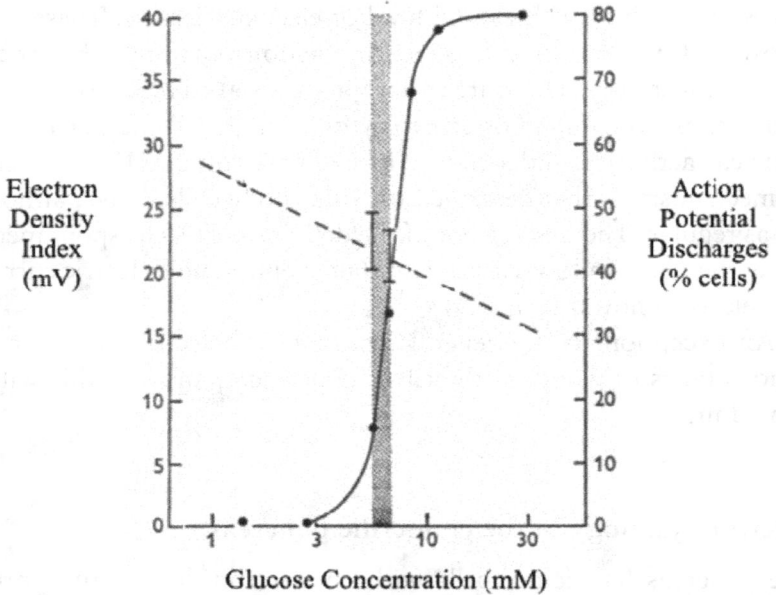

Figure 57: The electrophysiology of the islets of Langerhans β cells. Electron density changes not previously published.

Electrophysiology of the thyroid follicular cells

These cells have a resting electron density of 60–70 mV, which can be manipulated up and down by injecting and withdrawing electrons via micro-electrodes. Studies of functional depolarisation appear to be lacking, but the hyperpolarisation that follows blockade of the sodium pump suggests the presence of NCX, and therefore of Ca^{2+} involvement following functional depolarisation. Thyroid hormone enhances Ca^{2+} current in the heart.

Electrophysiology of the parathyroid glands

These glands control extracellular Ca^{2+} of the body. At a value of 2.5 mM extracellular calcium (due to albumen binding, free $[Ca^{2+}]$ is ~ 1.2 mM), electron density is 20 mV. Exposure to low Ca^{2+} extracellular solutions (1.5 mM) causes rapid hyperpolarisation to a mean value of 50 mV. Depolarisation by increasing extracellular K^+ causes an increase in hormone secretion, but this seems irrelevant to normal function as extracellular K^+ concentration is within a normal range in healthy mammals.

Electrophysiology of the pineal gland

Isolated cultured pinealocytes have a resting electron density, as measured using amphotericin-perforated patches, of 46 ± 1 mV. Such pineal cells show a sustained depolarisation mediated by a Ca^{2+}-conducting non-selective cation channel. However, this was detected when an adrenergic stimulus was applied that caused an initial hyperpolarisation. The method used, perforated patch clamp recording, may not reliably reflect normal function.

Testosterone-secreting cells (Leydig cells)

According to Kawa (*1987*), in Leydig cells under voltage clamp, depolarisations from electron density of 50 mV evoked transient inward currents which were identified as Ca^{2+} current, since replacement of external Ca^{2+} with Mn^{2+} reversibly diminished the current. The Ca^{2+} channel may be activated by physiological changes in electron density, leading to an influx of Ca^{2+}. That testosterone secretion results from intracellular Ca^{2+} increase is suggested by the involvement of Ca^{2+}-induced Ca^{2+} release via ryanodine receptors of the endoplastic reticulum (Costa, *2009*). This system is similar to that of cardiac muscle.

Adrenal gland

The cells in the adrenal cortex secrete steroids such as cortisol and aldosterone, while those in the medulla secrete adrenaline. The electron density in cortical cells is in the range of 65–72 mV. Depolarisation to electron density less than 40 mV induces an inward Ca^{2+} current. Intracellular Ca^{2+} facilitates adrenocorticotropic hormone (ACTH) secretion from the hypothalamus. In the medullary cells, electron density is much lower, approximately 20–40 mV, and are part of the sympatho-adrenal system delivering catecholamines into the bloodstream.

CHAPTER 25

Electrophysiology of exocrine glands

These glands secrete their product into a duct that conducts it to function at another organ, e.g., pancreatic and salivary ducts. However, the electrophysiology seems to depend on the same principles. The method of investigation of depolarisations frequently involves changes in extracellular K^+, which is not the situation in the intact animal with a relatively constant value of $[K_o^+]$. There is often an influence of the sympathetic and/or parasympathetic end-terminal neurotransmitters which affect target cell membrane electrical impedances. As with endocrine gland cells, the most common secondary messenger appears to be Ca^{2+} that releases the contents of the secretary organelles containing those specialised molecules that effect the specialised functions.

Pancreatic acinar cells

These depolarise from an initial electron density of about 40 mV. The main stimulus to such depolarisation is acetylcholine, presumably by lowering cell membrane electrical impedance (Matthews, *1973*). The consequence is, again, a rise in intracellular Ca^{2+}, i.e., Ca^{2+} signalling as second messenger, which is the interaction that may impact the kinetics of IP3 (inositol triphosphate or inositol) production and fluid secretion from the pancreatic acinar cells into the duodenum (Yule, 2015).

Salivary acinar cells

These cells possess a resting electron density of about 69 mV and the activation appears similar to that of cells of more thoroughly investigated tissues, such as muscle and nerve.

Other exocrine glands

There seems no reason to doubt that the same mechanisms are likely in other exocrine glands such as sweat, mammary, lacrimal, sebaceous, prostate and mucous glands, all of which show electrical phenomena.

CHAPTER 26

Exceptions to any general model

Electrophysiology of the liver

Hepatocytes differ from glands in that Ca^{2+} is not a second messenger. Although the electron density is only about 40 mV, the cell membrane does not leak much and there appears to be no voltage-sensitive Ca^{2+} channel in isolated hepatocytes. Presumably the proteins on the outer leaflet of the cell membrane that are required for such a channel are genetically absent (Sawanobori, *1989*). At the present time it is not possible to give precise relationships between electric events and the many complex functions of the liver.

Electrophysiology of the kidney

The electron density of renal tubular cells is about 80 mV. This high electron density appears to be related to the great activity of the Na^+/K^+ ATPase in these cells, which have a major role in Na^+ excretion into the urine (Na^+ being excreted if in excess concentration in plasma). The Na^+/K^+ ATPase or sodium pump, as already explained, removes 3 Na^+ from the cell, i.e., 3 atoms each with a missing electron, in exchange for 2 K^+, i.e., 2 atoms each with a missing electron. Thus the cell gains one electron for each exchange that adds to the cytoplasmic electron density. Katz hopes that his work on this opens new possibilities for evaluating the role of Na^+/K^+ ATPase on discrete anatomic subdivisions of the functionally heterogeneous nephron (Katz, *1982*). At the present time it is not possible to give precise relationships between electric events and the many complex functions of kidneys.

Electrophysiology of the lung

The electron density of alveolar epithelial cells measured in primary culture is 40 mV, while reported values range from 27–63 mV. An electron outflow of approximately 30 mV would be required to produce an outward electrochemical gradient for protons and opening of an H^+ channel to control pH. This is important for the exhalation of carbon dioxide, which is acidic (DeCoursey, *2000*). The important function of oxygen uptake does not require electricity, as the oxygen diffuses from the alveolar gas into the red blood cells within the pulmonary capillaries. The red blood cells have an electron density of only 9 mV and lack a sodium pump.

Electrophysiology of the ear

Electron density has been detected in the interior of the cochlea, with discharge elicited by transient sounds. Electron density has also been detected in the endolymph. This is a complex system that cannot be explored in any detail here. In the afferent nerves of the semicircular canals, spike frequencies are modulated by stimulation that activates the hair cell receptor conductance. The relation between receptor current and transmitter release cannot be studied in the intact semicircular canal (Martini *et al.*, *2015*). This situation recalls the response of vascular endothelium to hair cells of the glycocalyx, thought possibly to cause depolarisation.

Electrophysiology of the eye — an exceptional exception

The electrophysiology of the eye is unique. Retinal rod and cone cells differ from most of those previously discussed, in that they show activation by an increase in electron density (photon to electron conversion is also found in clouds, plants and solar panels), instead of by depolarisation. Photons (quanta) can knock electrons out of atoms. There is also a complex cGMP-gated channel. It is claimed that there is a $Na^+/Ca^{2+},K^+$ exchanger which is thought to be required for efficient cone-mediated vision (Vinberg *et al.*, *2017*). In the dark (at rest), the photoreceptors are quite active, constantly releasing neurotransmitters. After absorption of a photon, the resulting hyperpolarisation decreases the amount of neurotransmitters released. This means that light actually turns off neurotransmitter receptors. Hyperpolarisation and neurotransmitter release are graded inversely according to light intensity. There are glutamatergic, GABAergic and glycinergic neurones in the retina. Retinal neurotransmitters have been reviewed by Wu & Maple (*1998*). The rods and cones are connected to horizontal and bipolar cells that connect to the retinal ganglion cells.

One wonders about how the eye can distinguish between the frequencies of the electromagnetic waves (oscillation frequency of photons) that allow perception of colour, and how that information can reach the brain. The frequency range varies with different species, allowing some to see in the dark using the infrared spectra.

CHAPTER 27

Central and autonomic nervous systems

When it comes to explaining depolarisation of nerve cell bodies, we return to the most common way in which cells are activated to perform the function that has been assigned to them by natural selection. Nerve cells in the brain are often activated by depolarisation due to exposure to serotonin, i.e., chemically mediated depolarisation.

The serotonin system is very complicated because, although the brain works with serotonin, it does not synthesise this chemical. The cells have to draw serotonin from the blood, into which the cells in the gut that synthesise serotonin secrete it. This property of brain cells is the well-known serotonin re-uptake mechanism, of which inhibition by drugs is widely used in the treatment of depression. Thus there is a possibility of serotonin in the blood being in excess of that required by the brain, so that it has to be buffered by the serotonin re-uptake mechanism of the platelets, which can store it in dense granules until the platelets are destroyed in the spleen and replaced by new ones. Noradrenaline and acetylcholine (which we met before, acting upon the sinus node pacemaker current) also act as neurotransmitters in the brain along with other major neurotransmitters such as glutamate and gamma-aminobutyric acid, the main excitatory and inhibitory neurotransmitters respectively, as well as neuromodulators such as dopamine. These molecules are used by the nervous system to transmit messages between neurones, or from neurones to muscles (e.g., from motor neurone to skeletal muscle). Communication between two neurones happens in the synaptic cleft (the small gap between the synapses of neurones). The whole electrophysiology of the central nervous system is very complicated and a description will not be attempted here.

Nevertheless, I think it is worthwhile to consider how breathing occurs. Like the heartbeat, it is a necessity of continued life. Whereas the heart can continue to function in the absence of innervation, breathing is generated in the brain, continues through the apparently unconscious phenomenon of sleep, but can be controlled voluntarily.

The automaticity, a rhythmic act, is produced by networks in the hindbrain (the pons and medulla). The neural networks direct intercostal and phrenic nerves, and produce pressure gradients that move air into and out of the lungs. The respiratory rhythm and the length of each phase of respiration are thought to be set by reciprocal stimulatory and inhibitory interconnection of these brain-stem neurones.

However, in most mammals, the rhythm is slowed by cutting or anaesthetising the vagus nerves. This is the result of removal of the Hering-Breuer inflation reflex, in which stretch receptors in the lungs transmit nerve impulses to the medulla that terminate inspiration. In humans, this only occurs when the lungs are stretched way above the normal inspiratory volume (Guz & Noble, 1970).

In an attempt to explore whether afferent information from the vagus, intercostal and phrenic nerves could be perceived, a number of experiments were performed in volunteers, including effects on breath-holding time and detection of added loads to breathing. The conclusion was that a drive to breathe from the vagus led to contractions of the respiratory muscles which gave the subject an unpleasant sensation that was suppressed voluntarily until the subject was forced to take a breath. This accounts for the lengthening of breath-holding time with vagal and phrenic nerve block and muscle paralysis.

There is also a Hering-Breuer deflation reflex that can be demonstrated in humans (Guz, 1971).

Pneumothorax is a condition in patients in which a lung comes away from the chest wall and deflates. It is treated by inserting a tube into the cavity between the lung and the chest wall. By putting the outer end of the tube under water, the lung can expand, the air

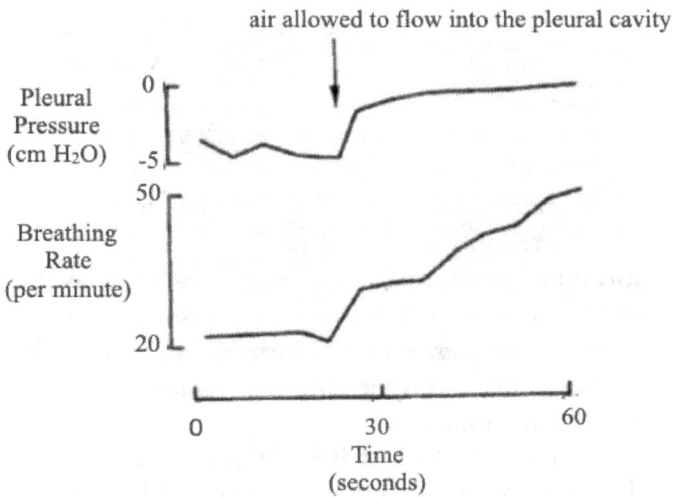

Figure 58:   Measurements made by the author during lung deflation in a patient with pneumothorax.

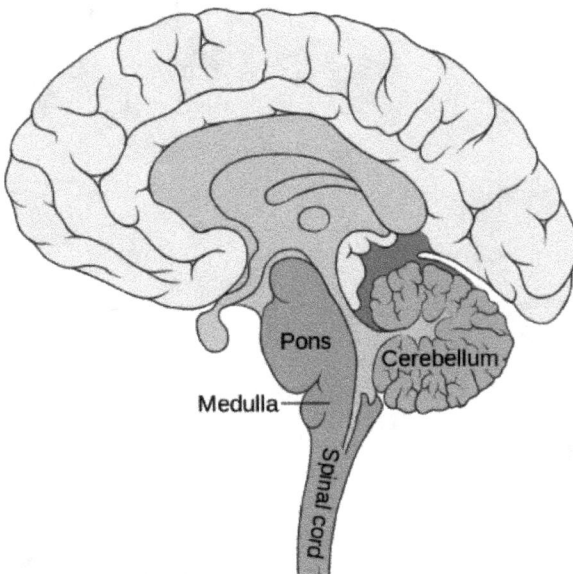

Figure 59:   Vertical section of the brain to show the location of the respiratory centre in the medulla. Licenced under Creative Commons Attribution-Share Alike 2.5.

coming out and bubbling out of the water trap. Disconnection of the tube allows the lung to deflate again causing more rapid breathing (Figure 58). This reflex serves to shorten exhalation when the lung is deflated. There is controversy over whether the reflex occurs by stimulation of stretch receptors or stimulation of proprioceptors activated by lung deflation. Like the inflation reflex, impulses from these receptors travel afferently via the vagus.

An important observation in experimental animals is that breathing continues after decerebration, e.g., Hayashi & Sinclair (1991). This confirms that the respiratory centre is in the hind brain (Figure 59). Breathing also continues in head injury cases in whom the cerebrum is non-functional.

The details of the control of breathing have been studied using electrodes inserted into the ventral medulla and by voltage imaging (Ikeda, 2017).

CHAPTER 28

Receptors affecting perception

Juxtacapillary or "J" receptors are located in the alveolar walls in close proximity to the capillaries. Because of their location, these receptors respond readily to chemicals in the pulmonary circulation, distention of the pulmonary capillary walls, and accumulation of interstitial fluid. These receptors were discovered by Dr. A.S. Paintal, and are also called pulmonary C receptors. Paintal (*1973*) thought that the increase in C fibre activity is a consequence of the rise in pulmonary artery pressure leading to a rise in pulmonary capillary pressure. This causes an increase in interstitial volume leading to excitation of the endings. His group postulated that the endings were located in collagen tissue which acts like a sponge.

Electron-microscopic evidence was obtained showing the presence of non-medullated sensory fibres in this collagen tissue (Paintal, *1973*). The afferent traffic runs in non-myelinated fibres in the vagus nerve. It has been postulated that this activity may only occur in pathological conditions of the lung and be perceived as dyspnoea.

Pain receptors

These are also called nociceptors, which are responsible for detecting harmful or noxious stimuli and transmitting electrical signals to the central nervous system. They are present in skin, viscera, muscles, joints and meninges to detect a range of stimuli, which may be mechanical, thermal or chemical in nature.

These receptors are unusual neurones because they have cell bodies with peripheral axons and terminals that respond to the stimuli, and central branches that carry the information into the central nervous system via non-myelinated nerve fibres (similar to J receptors afferent fibres). I suggest that the receptors transduce the change in electron density produced by the stimuli (postulated to be due to a decrease in cell membrane electrical impedance) into an action potential or, more commonly, a set of action potentials that encode the intensity of a noxious stimulus applied within their receptive fields.

The variety of cell types that are postulated to detect various noxious stimuli are represented diagrammatically in Figure 60. This variety may account for the perception of different kinds of pain that doctors ask patients about, e.g., burning, sharp, dull, etc.

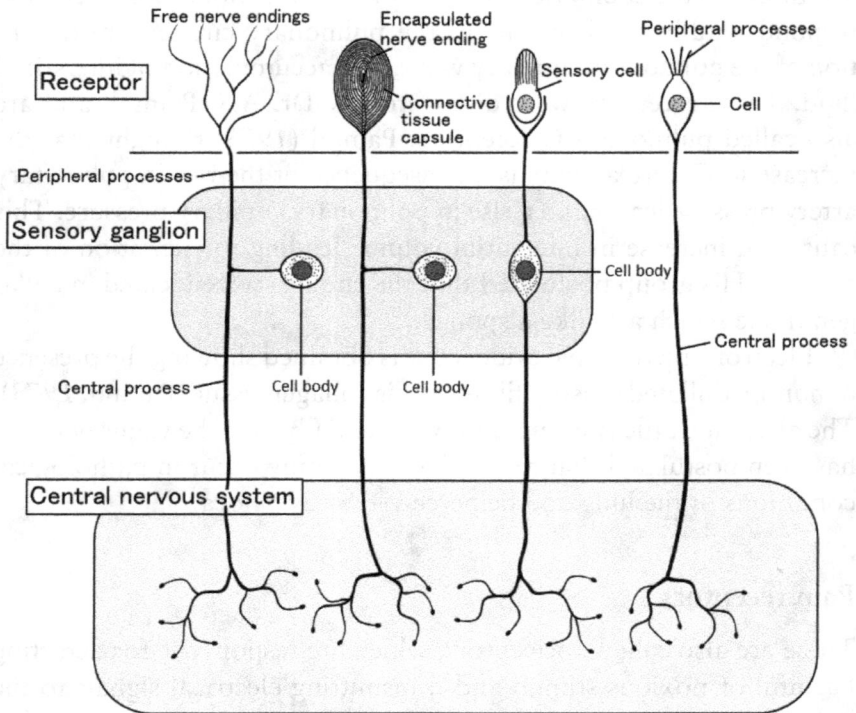

Figure 60: Complex of a variety of receptor neurones that sense pain. Licenced under Creative Commons Attribution-Share Alike 3.0.

Pain is a large subject in itself, so I suggest that the reader particularly interested in this subject consult Cech & Martin (*2012*).

Stretch receptors

The receptors that signal lung inflation in the Hering-Breuer inflation reflex are located in the bronchi and are examples of receptors present in many other locations in the body. Such stretch receptors are mechanoreceptors responsive to distention of various organs and muscles, and are neurologically linked to the medulla in the brain stem via afferent nerve fibres. Examples include stretch receptors in the arm and leg muscles and tendons, in the heart, in the colon wall, as well as in the lungs and bladder.

Stretch receptors are also found around the carotid artery, where they monitor blood pressure and stimulate the release of antidiuretic hormone from the posterior pituitary gland.

As far as I can ascertain from the literature, stretch receptors produce electron outflow from the afferent nerve endings. The depolarisations so induced are sufficient to trigger action potentials. Action potentials are usually larger in the slowly adapting than in the rapidly adapting neurone. The distributions of action potential amplitude are not sharply delimited, and there is an overlap from the two groups of neurones (Nakajima & Onodera, *1969*). There are no marked differences in the current/voltage relationship between the two types to account for the variations in speed of adaptation, but I would like to postulate that this could be due to the speed of electron restoration by the mitochondria, which, in turn, might be due to differences in mitochondrial membrane electrical impedance.

Muscle stretch receptors

The perception of position of the parts of the body is very important and is attributed to muscle spindles, which are stretch receptors within the body of a muscle that primarily detect changes in its length (Figure 61, labelled Organ of Golgi but not to be confused with Golgi tendon organs — see Figure 62). They convey length information to the central nervous system via Ia afferent nerve fibres. This information can be processed by the brain as proprioception. Also claimed as contributing to proprioception are the Golgi tendon organs that sense

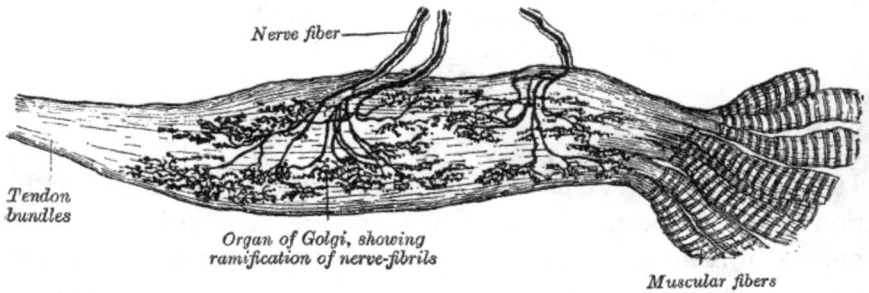

Figure 61: Right: striated muscle ending. Left: tendon. Centre: Golgi stretch receptor consisting of a fibrous network that stimulates the nerve endings directly, resulting in action potential afferent traffic from which a contribution to perception of proprioception is a widespread opinion. Source: Wikimedia Commons, public domain.

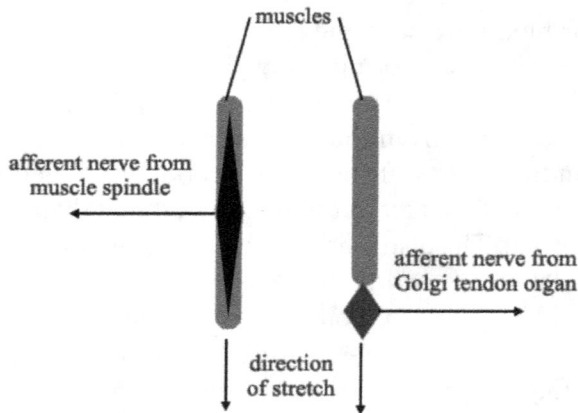

Figure 62: Left: muscle spindle (black) is in parallel with muscle. Right: Golgi tendon receptor (dark grey) in series with muscle.

changes in muscle tension. It lies at the origins and insertions (Moore, *1984*) of skeletal muscle fibres into the tendons of skeletal muscle.

Golgi tendon organs are in series with the muscle fibres. When a muscle is passively stretched, most of the change in length occurs in the muscle fibres, since they are more elastic than the fibrils of the tendon. When a muscle actively contracts, however, the force acts directly on the tendon, leading an increase in the tension of the collagen fibrils in the tendon organ and compression of the intertwined sensory receptors. As a result, Golgi tendon organs are exquisitely sensitive to increases in muscle tension that arise from muscle contraction but, unlike spindles, are relatively insensitive to passive stretch. Each Golgi tendon organ creates afferent traffic in Ib axons, which are slightly smaller than Ia axons to which afferent traffic from muscle spindles relate.

The stretch reflex is a muscle contraction in response to stretching within the muscle. The reflex functions to maintain the muscle at a constant length.

Readers who have been examined by doctors will be familiar with the stretch reflex in which a knee tap elicits a muscle contraction moving the lower leg forward. This will not be elicited if there is a fault in the stretch receptors, a fault in the sensory nerve which carries the nerve impulses from the stretch receptors to the spinal cord, a fault in the spinal cord circuit, a fault in the motor nerve, or a fault in the quadriceps muscle.

The example of an upper limb circuit is illustrated in Figure 63.

Gut stretch receptors

The extensive web of nerve endings lining the gut plays an important role in controlling how much the subject eats by monitoring the contents of the stomach and intestine and then sending signals back to the brain. These may cause perceptions that boost or lower appetite. Many scientists also postulate that this feedback involves hormone-sensitive nerve endings in the gut that track the nutrients consumed, but no one has yet tracked down the exact types of neurones that convey such signals to the brain. Actually, the regulation of food intake in the gastrointestinal tract is poorly understood. The vagus nerve monitors gastrointestinal signals, and vagal afferents consist of a diverse array of neurones whose sensory endings are thought to be specialised for detection of particular stimuli.

There have been electrophysiological and morphological analyses of vagal mechanoreceptors in the gut wall, with conflicting conclusions (Phillips & Powley, 2000). Electrophysiologists have distinguished a single general class of ending in smooth muscle, one characterised as an 'in series' tension receptor. Morphology, in contrast, has characterised two distinct specialisations of vagal afferent endings in the muscle wall of the gastrointestinal tract. These two structures differ in terms of their target tissues, terminal architectures

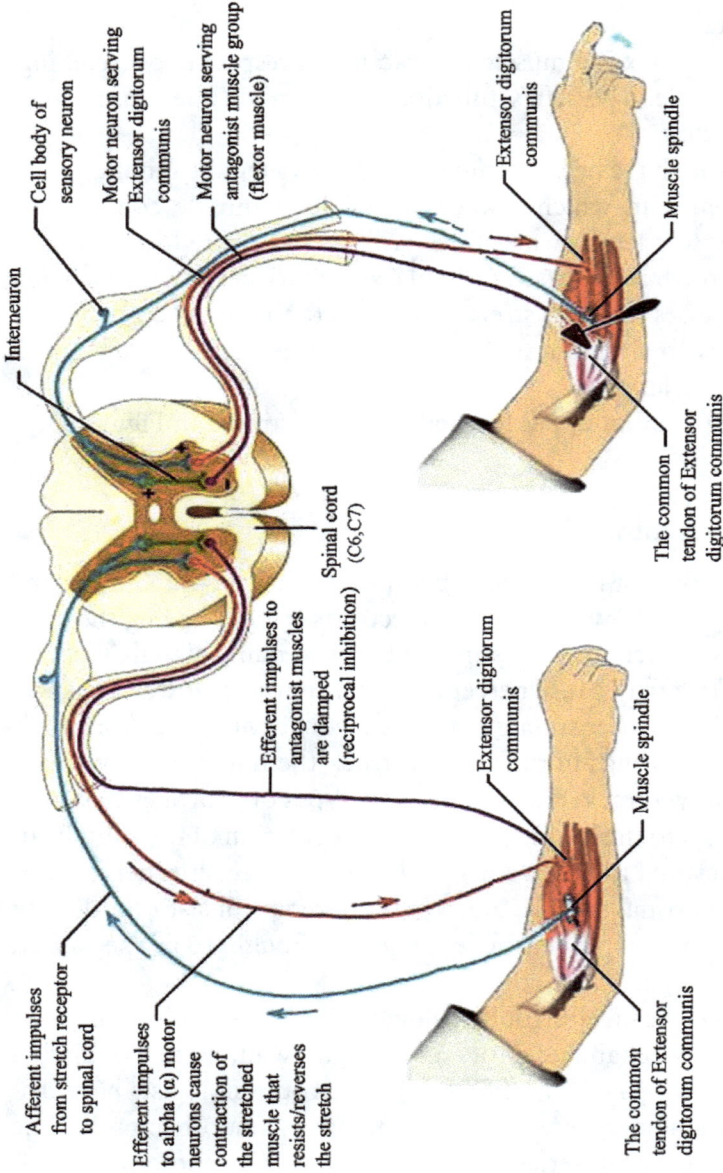

Figure 63: Another stretch reflex. The extensor digitorum reflex is an example of a monosynaptic reflex (also called the biceps reflex or brachioradialis reflex). The afference limb involves receptor (muscle spindle in extensor digitoium), afferent (Iα fibre), and center (spinal cord C6, C7). The efferent limb involves alpha motor neurons. The effector organ is a skeletal muscle, the extensor digitoium. The afferent limb passes to the efference limb via the spinal segments C5 and C6. Licenced under Creative Commons Attribution-Share Alike 4.0.

and regional distributions; they also develop on different ontogenetic timetables and depend on different trophic support in the muscle wall. On the basis of these features, it has been proposed that one of the putative mechanoreceptors, the intra-ganglionic laminar ending, has characteristics of a tension receptor and the other, the intramuscular array, has features of a stretch or length receptor, in a functional analogy with striated muscle proprioceptors. The former should have similarities to Golgi tendon organs, whereas the latter should have similarity with muscle spindle afferents.

Touch receptors

This is a subtype of sensory neurone that is located in the skin, a specialised ending that responds to mechanical stimulation. It is thought that the first step in the activation of the receptor is stretch of the cell membrane. I would expect this to lead to a decrease in cell membrane impedance, leading to an outflow of electrons causing depolarisation and opening of ion channels including the Ca^{2+} channel. The resultant total depolarisations are sufficient to produce action potentials that are transmitted to the central nervous system, leading to perception by the brain of the awareness of touch.

There are four major types of encapsulated mechanoreceptors (Figure 64) that are specialised to provide information to the central

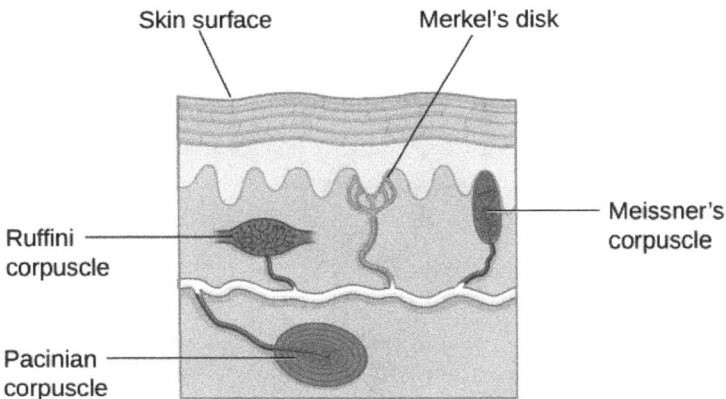

Figure 64:   Touch receptors. Licenced under Creative Commons Attribution Share-Alike 4.0.

nervous system about touch, pressure, vibration, and cutaneous tension.

Merkel's receptor occurs in the outer layer (epidermis) of hairless skin that responds to sustained pressure. Named after German anatomist Friedrich S. Merkel (1845–1919), these discs are found in large numbers in the fingertips, where they are important in discriminating textures and shapes. As part of the somatosensory system, touch receptors can transmit information to the central nervous system resulting in the perception of touch.

Ruffini corpuscles are enlarged dendritic endings with elongated capsules that encapsulate mechanoreceptors which detect skin stretch and deformations within joints. I postulate that the stretch of the cell membranes reduces cell membrane impedance, allowing outflow of electrons and generation of action potentials. Their feedback to the central nervous system helps the brain to coordinate the gripping of objects and controlling of finger position and movement. They are also claimed to contribute to proprioception (kinesthesia). Ruffini endings are also thought to detect warmth.

Meissner's corpuscles (Figure 65) respond to touch and low-frequency vibration, being closely located under the epidermis (Figure 64). Pacinian corpuscles are also mechanoreceptors; they respond to pressure applied to the skin's surface, being the deepest of the touch receptors (Figure 64).

Figure 65:   Meissner's corpuscle. Diagram representing a pressure receptor of the skin responsible for sensitive sense of touch (light pressure). Licenced under Creative Commons Attribution-Share Alike 3.0.

Somatosensory perception involves the activation of primary sensory neurones whose cell bodies reside within dorsal root ganglia and cranial sensory ganglia. Dorsal root ganglia neurones are pseudo-unipolar, with one axonal branch that extends to the periphery and associates with peripheral targets, and another branch that penetrates the spinal cord and forms synapses upon second order neurones in the spinal cord grey matter and, in some cases, the dorsal column nuclei of the brainstem.

CHAPTER 29

Receptors initiating feedback control reflexes

Baroreceptors

These are essentially stretch receptors that detect changes in blood pressure at two key sites in the aortic arch and the carotid sinus. Information from the carotid baroreceptors appears to be more important, and sensory fibres from here are carried in the glossopharyngeal nerve to the brainstem. Sensory fibres from the aortic arch are carried by the vagus nerve, which is also a conduit for parasympathetic efferent fibres to the heart (Figures 66 and 67). Changes in beat-to-beat arterial pressure pulses are processed in the medulla oblongata where alterations to sympathetic and parasympathetic activity are made.

They are responsible for a medullary reflex that responds to a rise in arterial pressure (so-called blood pressure or BP) to cause a corrective response (negative feedback) from peripheral resistance vessels and heart contractile force.

Baroreflex

Afferent nerve traffic from carotid baroreceptors travels to the medulla of the hind brain by cranial nerve IX (glossopharyngeal). Afferent traffic from aortic baroreceptors travels to the medulla by cranial nerve X (vagus). The efferent pathway goes first to the sympathetic ganglion

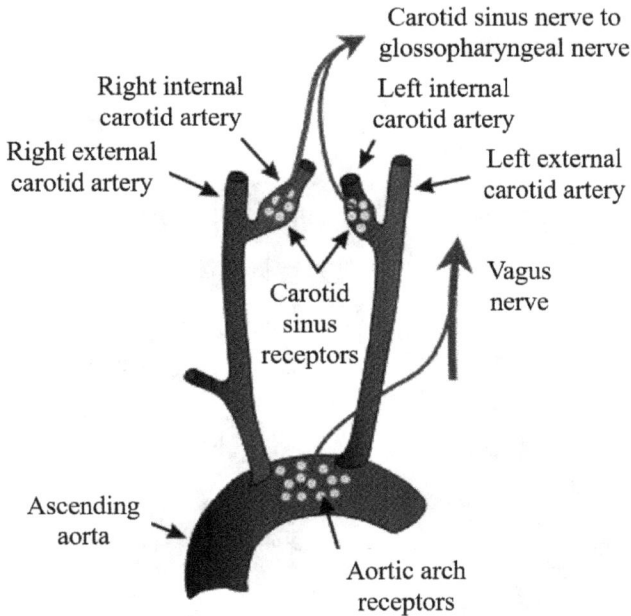

Figure 66:   Location and innervation of baroreceptors. Original drawing by CV Physiology (reproduced with permission from CV Physiology), re-labelled by author.

with acetylcholine as neurotransmitter, then to the efferent postganglionic nerves to heart and peripheral vessels, releasing noradrenaline.

Taste and smell receptors

Smell and taste receptors are chemoreceptors that sense the chemical environment. There are specialised taste receptors within the mouth that are referred to as taste cells and are bundled together to form taste buds. The large taste cells of the aquatic salamander Necturus are electrically excitable and generate action potentials (Roper, 1983) in response to electron outflow. Vertebrate species regularly generate action potentials not only on electrical stimulation, but also in response to apically applied chemical stimuli. Taste cells appear to be the only non-neuronal sensory receptor cells to generate action potentials.

Conventionally there are said to be 4 types of taste buds that detect sweet, sour, salty and unami stimuli (savouriness of protein).

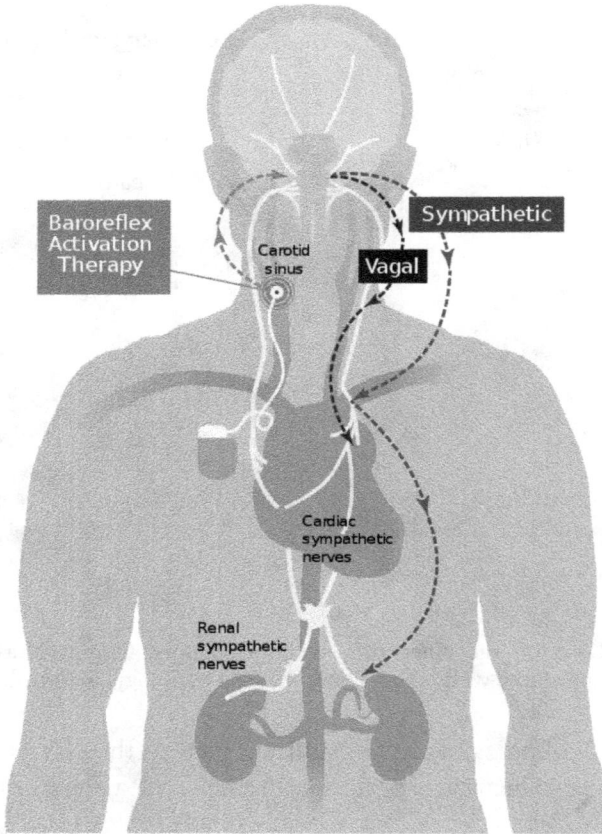

Figure 67: The baroreflex. Licenced under Creative Commons Attribution-Share Alike 4.0.

However, there are probably more stimuli (like smell odour stimulants). As we have sour, bitter seems to me almost the same, so why not savoury? There are a large number of combinations of molecules in the air (dots above the taste bud labelled A) in Figures 68 and 69 that will result in a wide variety of tastes perceived.

Molecules in the air stimulate receptors in the nose (Figure 70). If the organism has no receptor for a particular odour, it will not be detected. The senses of smell and taste are directly related as they have similar receptor types.

Non-function will affect both smell and taste and will not be detected by the individual.

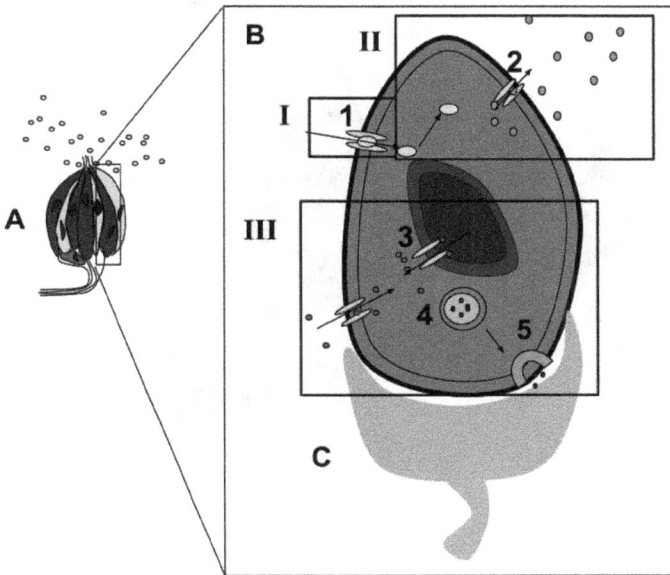

Figure 68: (Conventional interpretation) The diagram depicts the signal transduction pathway of the sour taste. Object A is a taste bud, object B is a taste receptor cell within object A, and object C is the neuron attached to object B. Part I is the reception of hydrogen ions or sodium ions. 1. If the taste is sour, H^+ ions, from an acidic substance, pass through their specific ion channel. Some can go through the Na^+ channels. If the taste is salty, Na^+ ions pass through the Na^+ channels. Electron outflow takes place. Part II is the transduction pathway of the relay molecules. 2. Cation channels, such as K^+, are opened. Part III is the response of the cell. 3. An influx of Ca^+ ions is activated. 4. Ca^+ activates neurotransmitters. 5. A signal is sent to the neuron attached to the taste bud. Licenced under Creative Commons Attribution-Share Alike 4.0.

Other chemoreceptors

Chemoreceptors are located in the brain stem and in peripheral arteries. The peripheral chemoreceptors, located in the carotid bodies, respond primarily to low oxygen saturation of the blood (hypoxaemia) usually resulting from poor oxygen uptake in the lungs (hypoxia). They are activated by changes in the partial pressure of oxygen (P_aO_2) and trigger respiratory drive changes aimed at maintaining normal partial pressure levels. Central chemoreceptors located in the region of the brainstem respond to increases above normal in the partial pressure of carbon dioxide (P_aCO_2) in the blood (hypercapnia). Activation of either the hypoxic or hypercapnic chemoreflex elicits

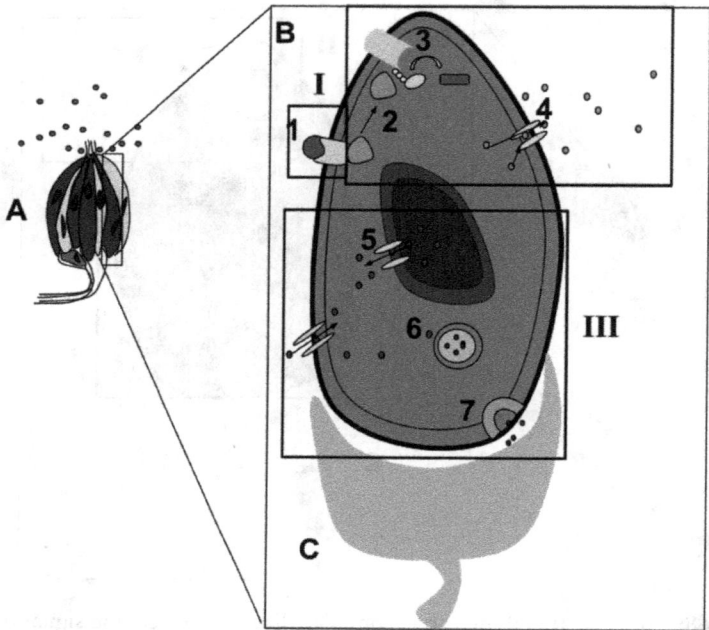

Figure 69: (Conventional interpretation) The diagram above depicts the signal transduction pathway of the sweet taste. Object A is a taste bud, object B is one taste cell of the taste bud, and object C is the neuron attached to the taste cell. Part I shows the reception of a molecule. 1. Sugar, the first messenger, binds to a protein receptor on the cell membrane. Then there is transduction of the relay molecules. 2. G Protein-coupled receptors, second messengers, are activated. 3. G Proteins activate adenylate cyclase, an enzyme, which increases the cAMP concentration. Electron outflow occurs. 4. The energy, from step 3, is given to activate the K^+ protein channels. Part III shows the response of the taste cell. 5. Ca^{2+} protein channels are activated. 6. The increased Ca^{2+} concentration activates neurotransmitter vesicles. 7. The neuron connected to the taste bud is stimulated by the neurotransmitters. Licenced under Creative Commons Attribution-Share Alike 4.0.

both hyperventilation and sympathetic activation. Apnoea (cessation of breathing) is often recorded transiently during sleep (sleep apnoea). When the inhibitory influence of stretch of the pulmonary afferents is eliminated, there is a potentiation of the sympathetic response to both hypoxia and hypercapnia. This inhibitory influence of the pulmonary afferents is more marked in the sympathetic response to peripheral compared with central chemoreceptor activation. The arterial baroreflexes also have a powerful inhibitory influence on the chemoreflexes.

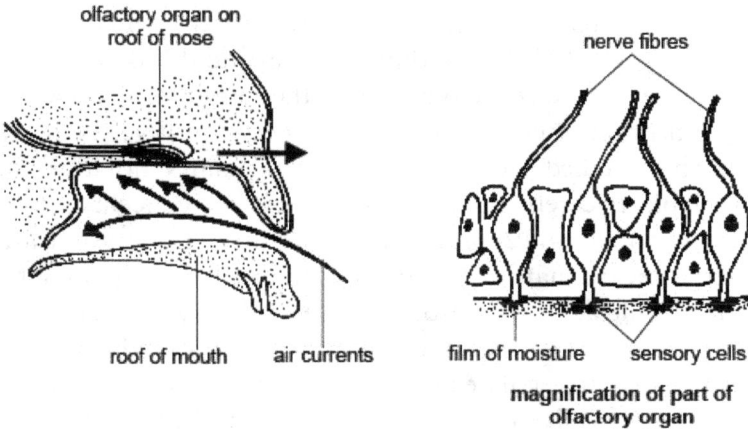

Figure 70:   Left: Afferent pathway for smell sense from the nose and tongue (air currents carry odorous molecules). Right: Sensory cells subject to electron outflow create action potentials in the nerve fibres. Licenced under Creative Commons Attribution-Share Alike 3.0.

This inhibition is again more marked with respect to the peripheral compared with central chemoreflexes.

Hallowill (*2003*) assessed baroreflex control of heart rate and muscle sympathetic nerve activity during normoxia, hyperpnoea at constant P_aCO_2, and isocapnic hypoxia. While isocapnic hyperpnoea did not alter heart rate, arterial pressure, or sympathetic outflow, hypoxia increased heart rate, mean arterial pressure and sympathetic activity. The sensitivity for baroreflex control of both heart rate and sympathetic activity was not altered by either hypoxia or hyperpnoea. The authors thought that acute activation of peripheral chemoreceptors with isocapnic hypoxia resets baroreflex control of both heart rate and sympathetic activity to higher pressures without changes in baroreflex sensitivity. These effects appeared to be largely independent of breathing rate and tidal volume.

I have not found data on the mechanism by which hypoxaemia, for instance, creates changes that produce action potentials.

The traditional concept of the function of central chemoreception is that it, along with peripheral chemoreception at the carotid body, 1) regulates arterial P_aCO_2 within normal limits in response to primary changes in CO_2, and 2) regulates blood and body pH in response to acid-base disturbances. Central chemoreception refers to the detection of changes in CO_2/H^+ within the brain and the associated effects on breathing. In the conscious animal the response of ventilation to changes in brain interstitial fluid pH is very sensitive (Nattie & Li, 2012). They studied the relationship of alveolar ventilation (dependent variable) to cerebrospinal fluid hydrogen ion concentration (independent variable) and found it to be steeply rising and not affected by arterial bicarbonate concentration, in a conscious goat subjected to both chronic acid-base disorders and acute CO_2 inhalation. These authors speculate that central chemoreception responds to small variations in P_aCO_2 to regulate normal gas exchange and to large changes in P_aCO_2 to minimise acid-base changes. Central chemoreceptor sites

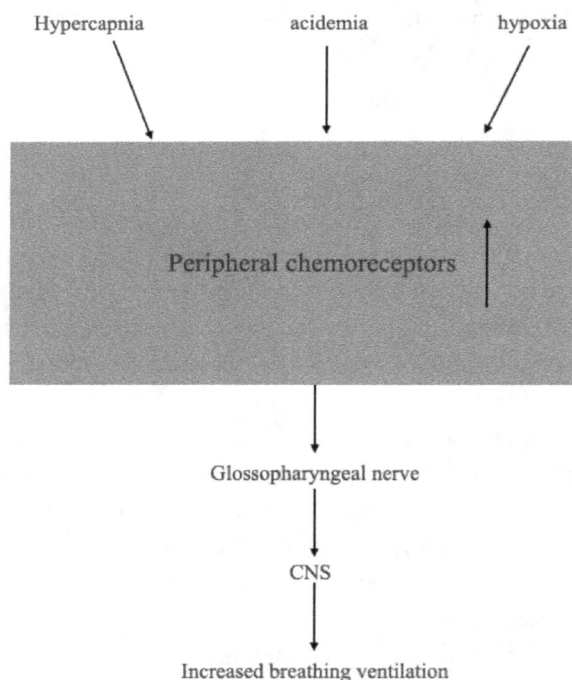

Figure 71: The chemoreflex complex. Peripheral chemoreceptor stimulation causes afferent neural traffic in the glossopharyngeal nerve, resulting in activation of the dorsal root ganglion (DRG) and an increase in breathing ventilation. A rise in P_aCO_2, accompanied by increased acidity (fall in pH), stimulates peripheral chemoreceptors but also, through a direct effect in the brain medulla, stimulates ventilation.

vary in function with sex and development. From an evolutionary perspective, central chemoreception may have grown out of the demands posed by air vs. water breathing, homeothermy, sleep, optimisation of the work of breathing with the 'ideal' arterial P_aCO_2, and the maintenance of the appropriate pH at 37°C for optimal protein structure and function.

The interaction between central hindbrains and peripheral chemoreception is illustrated in Figure 71.

In 1974, there was great temporary excitement when my group thought they had discovered a new reflex elicited from inhalation of carbon dioxide into the lung during closed chest cardiopulmonary bypass (Bartoli, *1974*). This enthusiasm was somewhat diluted when it was discovered that, in the abnormal composition of lung gases, under conditions of cardiopulmonary bypass, the effect was caused by a change in sensitivity on the stretch receptors (Bradley, *1976*). Thus, there are no carbon dioxide receptors in the lungs that stimulate ventilation under normal conditions.

The lesson I learned from this is that if your experimental preparation alters normality, misleading results will be obtained.

The more interference with normality of your experiment, the more likely that wrong conclusions could emerge, e.g., isolated squid axon with interior milieu removed, a bit of patch-clamped cell membrane.

CHAPTER 30

Summary and general comments on vertebrate animals

Figure 72 is a schema which seems to fit, with variations at each step, the electrophysiology of many types of cell such as nerve (except axons), muscle, endocrine and exocrine glands. From the negative membrane potentials in the conventional scheme, I derive an electron density index which is positive and I postulate that electrons flow (i.e., there is electricity) from higher to lower electron density sites, taking into account the intervening electrical impedance. It also takes into account the lack of correlation of membrane potential and electrolyte distribution comparing different tissues. Exceptions from this general scheme occur in liver, lung, kidney, eye and ear, all of which show different kinds of electrical events. Ionic flow also occurs causing currents of which the Ca^{2+} current seems to be the most important, but both that and any other ionic currents create the problem of possible ionic imbalance, which has to be avoided so that ions in equal ions out in the steady state.

I have excluded from the diagram the corresponding scheme for the afferent receptor system, because of the lack of data on the detailed electrical events in this system other than the final results — electric afferent travel in afferent nerves and the putative neuronal connections.

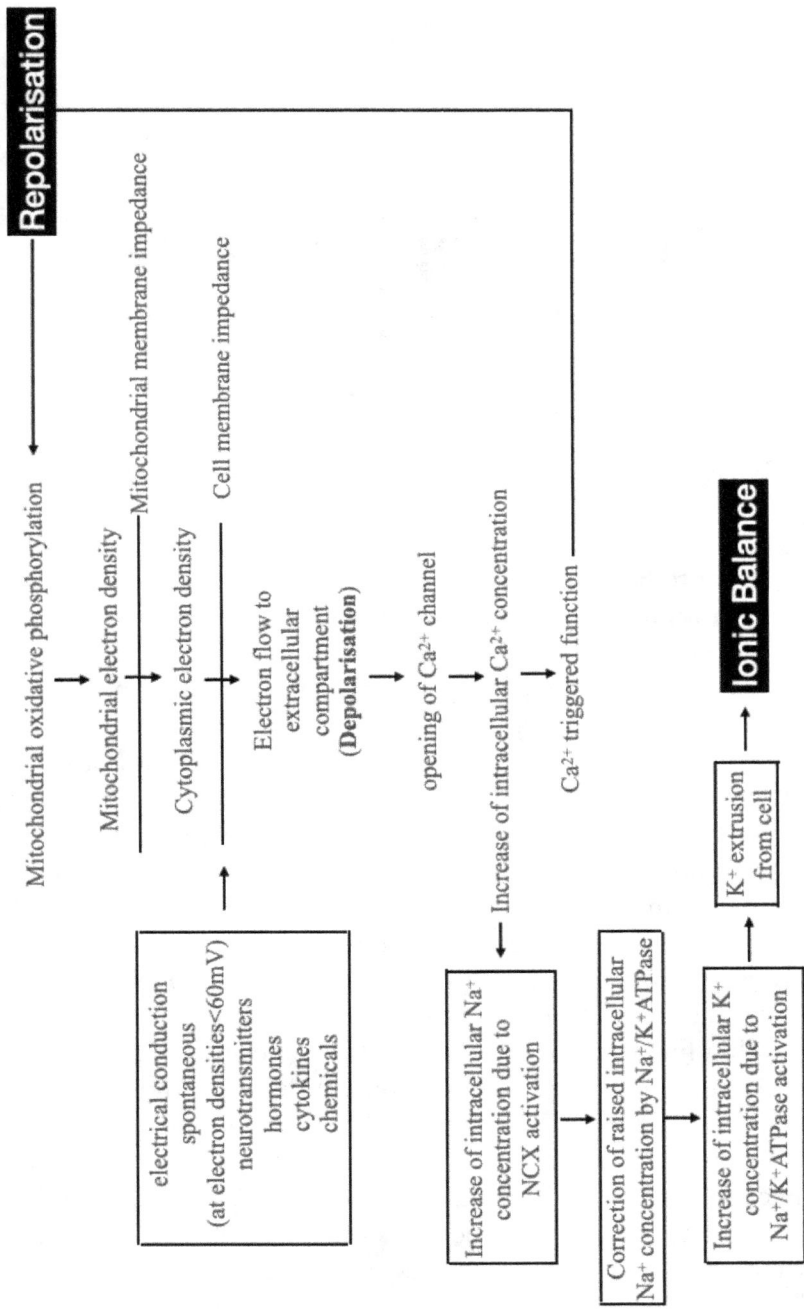

Figure 72:   Schema of a system of electrophysiology of vertebrate motor cells that fits many different tissue types.

CHAPTER 31

The electrophysiology of invertebrates

Insects

Insects appear to be the most successful of the members of the animal kingdom without vertebrae and are the only group of invertebrates that have evolved wings and flight. Two insect groups, the dragonflies and the mayflies, have flight muscles attached directly to the wings. In other winged insects, flight muscles attach to the thorax, which make it oscillate in order to induce the wings to beat. Of these insects, some achieve very high wingbeat frequencies (up to 200 wingbeats per second!) through the evolution of a system in which the thorax oscillation is faster than the rate of nerve impulses, e.g., insects that beat their wings more rapidly, such as the bumblebee with muscle that contracts more than once per nerve impulse.

How is this oscillating wing beating possible? To my knowledge, only the Iwazumi theory gives us a clue. It is postulated that the actin thin filament tips are pulled into the position illustrated in Figure 41. That causes a pull on the connection to the thorax which reacts mechanically by pulling the tip back out of the field; then the field pulls the tip back again. The combination of field force and mechanical arrangement thus has a resonant frequency that determines the wing oscillation frequency, so that the oscillation continues in the absence of nerve impulses as long as the field is maintained. This has been called a classic stretch activation and shortening deactivation phenomenon, but in Iwazumi's scheme the activation is the result of an increase in quanta of the field. The frequency of this oscillation is independent of the strength and frequency of electrical stimulation and is not associated with action potentials in synchrony with the oscillation (Ruegg, 1967).

Ruegg showed that the activity of the synchronous insect flight muscle oscillation with its myogenic rhythm can be increased by raising the concentration of calcium ions. Nowadays, I would state that the field develops as a result of Ca^{2+} increase; the isometric force-pCa relation ($pCa = -\log_{10}$ of the calcium concentration) shows that 1.5–2 units of pCa are necessary to raise isometric force from its threshold ($pCa \sim 6.5$) to its maximum (Linary, 2004).

Ruegg, who studied isolated myofibrils, also stated, "Since the functionally isolated myofibrillar system, consisting essentially of actin and myosin filaments, contracts and relaxes during oscillation up to a million times with ATP as the only energy source and at a constant (i.e., buffered) level of ionised calcium, it seems that in the living cell, too, oscillation is ATP-driven and dependent on the presence of calcium ions but independent of the rapid "give and take" of calcium ions by the sarcoplasmic reticulum (SR). Thus, during flight, little energy is wasted for the calcium-pumping action of the SR." Between 10^{-7} and 10^{-6} M $[Ca^{2+}]$, the increase of delayed tension and oscillatory extra ATPase activity is almost proportional to the logarithm of the calcium concentration. The mechano-chemical coefficient relating work and consumption of ATP is about 3,000 cal per mol of ATP split.

Figures 73 and 74 give convincing evidence that basic mechanisms (in this case actin filaments sliding to and fro within an electromagnetic field in muscle) are similar to those in vertebrates. Perhaps this is not surprising as vertebrates probably evolved from invertebrates during the space-time of evolution. Figure 74 illustrates wing oscillation.

Many invertebrates, including flies, cannot keep themselves warm in cold environments ("cold-blooded"), whereas many vertebrates can do so ("warm-blooded"). Some flies are cold-resistant while others are not, and this difference seems to be related to differences in electron density. The more cold-tolerant species are better able to defend

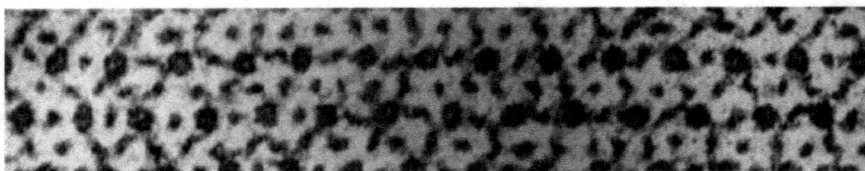

Figure 73: Electron micrograph of oblique section of insect flight muscle. Note the complete regularity of the quantum field of myosin thick filaments (large dots in cross section, some showing projections), each surrounded by thin actin filaments (small dots in cross section). Reproduced with permission from Iwazumi (deceased) before his death.

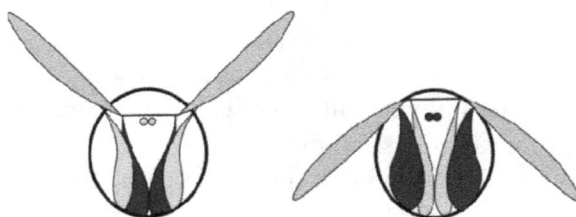

Figure 74: Insect wing oscillation. Licenced under Creative Commons Attribution-Share Alike 3.0.

their electron density at lower temperatures. The variation in lowest tolerable temperature among species was not explained by a higher electron density of muscle at 20°C. One of the most cold-tolerant species (*D. persimilis*) had the least electron density of all the species at 20°C (55.5 mV). Greater baseline electron density is clearly not a general strategy among all cold-tolerant drosophilids. Maybe other cold-tolerant drosophilids can maintain electron density across a broad range of temperatures, as *D. persimilis* does. However, most insects, e.g., cockroaches and honey bees, show cold-induced neuromuscular impairment, associated with chill coma and linked to a decrease of excitability due to an electron outflow of the central nervous system and muscle (Anderson, *2015*).

Molluscs

Molluscs are a large invertebrate phylum which includes snails, slugs, mussels, clams, squid and octopuses. They have a soft unsegmented body and live in aquatic or damp habitats, and most kinds have an external calcareous shell. Humans tend to destroy slugs and snails (they are different species) because they eat garden plants. However, slugs and snails are very important. They provide food for all sorts of mammals, birds, slow worms, earthworms, insects and they are part of the natural balance. Upset that balance by removing them and we can do a lot of harm. Thrushes, in particular, thrive on them! Slugs, along with earthworms, help to keep soil fertile in organic farms, another reason not to destroy them.

Slugs

Slugs (Figure 75) have light and dark bands which represent oblique muscle fibres, relaxing and contracting in what are technically known as pedal waves. There are 2 sets of these muscle fibres with different functions. To move forwards, the fibres directed inwards and rearward contract between waves, pulling the slug from the front and pushing off toward the back. Simultaneously, the fibres directed inward and forward pull the outer surface of the sole forward, generating each pedal wave (Gordon, 2010).

The mechanism of contraction is by activation of the nicotinic acetylcholine receptor of the muscles, then there is depolarisation (electron outflow) that generates an action potential of the membrane. This is followed by an interaction between actin and myosin. This is very similar to the mechanism in skeletal muscle of vertebrates.

Snails

Snails are like slugs but have a shell. The electrophysiology is similar to that of slugs.

Squid

Squid (Figure 76) are aquatic cephalopods in the superorder Decapodiformes with elongated bodies, large eyes, eight arms and two tentacles. Like all other cephalopods, squid have a distinct head, bilateral symmetry, and a mantle.

Slug

Keel (a ridge along the back) · Trunk · Respiratory pore (pneumostome) · Mantle · Optic tentacles with eyes at the tips · Head · Foot · Skirt · Anus (under the mantle) · Genital opening · Pedal slime gland · Mouth with file-like radula · Sensory tentacles

©EnchantedLearning.com

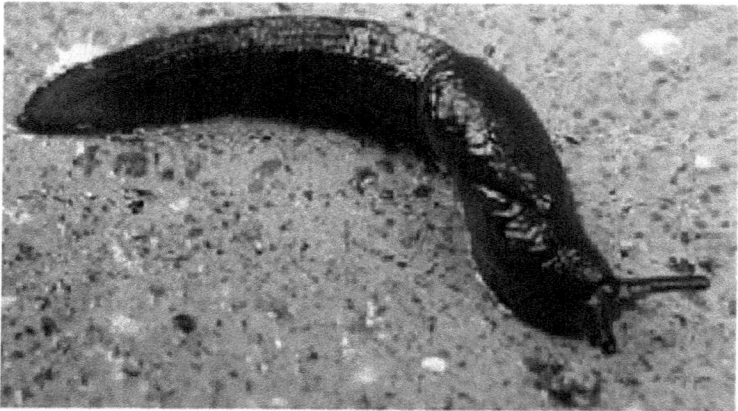

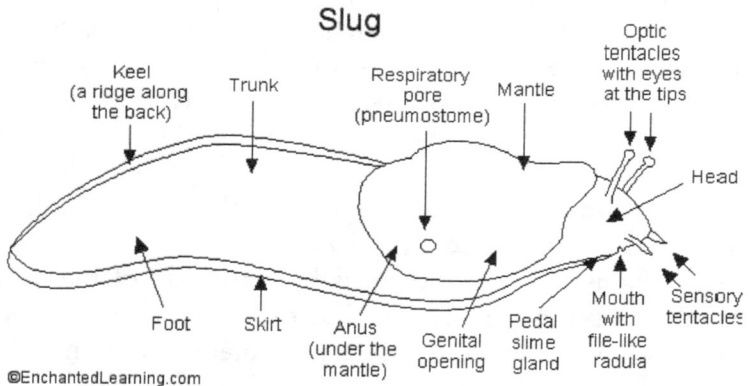

Figure 75: A slug. Licenced under Creative Commons Attribution-Share Alike 4.0.

The squid giant axon is well known to physiologists because of the classical studies of the Huxleys on the function of nerve. They chose this nerve because it is very large (up to 1.5 mm in diameter; typically around 0.5 mm), an axon that controls part of the water jet propulsion system in squid.

These great physiologists (Hugh and Andrew Huxley, not related) were able to access the lumen of the giant axon (Figure 77) and showed that a current passed inwards (conventional polarity) was dependent on the presence of sodium ions in the extracellular fluid and that a current passed outwards was dependent on the presence of potassium ions in the intracellular fluid (Hodgkin & Huxley, 1952).

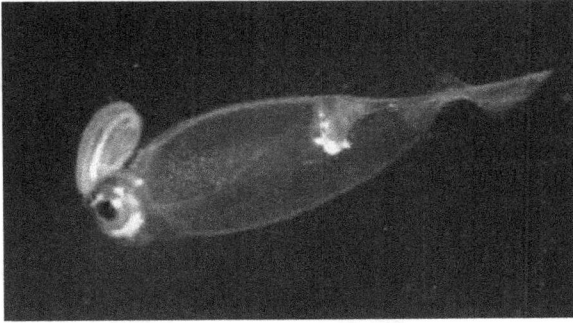

Figure 76:   A bathyscaphoid squid (*Teuthowenia megalops*) swimming in the "cock-atoo" posture. The animal rotates around the spindle-shaped digestive gland, which it keeps vertical regardless of the orientation of the mantle and head. Source: Wikimedia Commons, public domain.

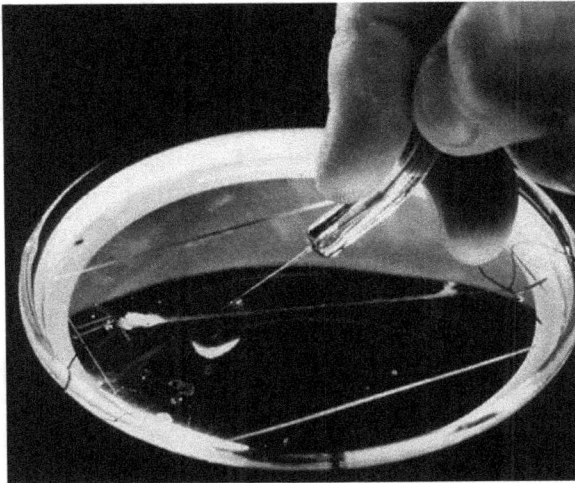

Figure 77:   The large size of the isolated squid axon is a convenient preparation for physiologists who can apply electrical stimulating and recording electrodes. NIH History Office from Bethesda — Giant Axon of Squid. Public Domain.

Unfortunately, they were ignorant of the fact that the intracellular compartment is not fluid, but a gel containing vital organelles such as mitochondria. These were all removed and replaced by a solution with the correct ionic composition. The unfortunate unforeseen consequence was that physiologists assumed that the electric charge of the inward current was carried by sodium ion flow. Not true — see the account in the section on the electrophysiology of the sinus node cell.

Unfortunately for the poor squid, it is a delicacy that is eaten in large quantities by humans. Fishing for squid, called squid jigging, uses underwater lights and takes place during the day but, depending on squid concentrations, also are used during the night and hauled up later on. Edible Argentine shortfin squid are found in the shallow part of the sea near a coast and overlying the continental shelf. They are found from the surface to about 800 metres in depth.

Octopus

Octopuses (Figure 78) are soft-bodied, eight-limbed aquatic molluscs. Around 300 species are recognised, and the order is grouped within the class Cephalopoda with squids, cuttlefish, and nautiloids. The octopus is bilaterally symmetric with two eyes and a beak, with its mouth at the centre point of the eight limbs. The soft body can rapidly alter its shape, enabling octopuses to squeeze through small gaps. They trail their eight appendages (tentacles), containing radial, oblique and longitudinal muscles, behind them as they swim. The siphon is used both for respiration and for locomotion, by expelling a

Figure 78: Common octopus (*Octopus vulgaris*). Licenced under Creative Commons Attribution-Share Alike 4.0.

jet of water. Octopuses have a complex nervous system and excellent sight, and are among the most intelligent and behaviourally diverse of all invertebrates.

Contraction of the radial muscles thins the wall, thereby increasing the enclosed volume of the sucker. If the sucker is sealed to a surface the cohesiveness of water resists this expansion. Thus, the pressure of the enclosed water decreases instead. The meridional and circular muscles antagonise the radial muscles. The muscle cells are excitable but have a broad range of linear electrical properties. They are electronically very compact so that localised synaptic inputs can control the electron density of the entire muscle cell. The octopus muscle action potential arising from electron outflow is primarily mediated by a calcium current through L-type voltage-gated channels. It is reasonable to assume that this channel, as in all muscle, is opened by electron outflow causing depolarisation.

Mussels

Mussel (Figure 79) is the common name used for members of several families of bivalve molluscs from saltwater and freshwater habitats. These groups have in common a shell whose outline is elongated and asymmetrical compared with other edible clams, which are often more or less rounded or oval.

Mytilus edulis is edible and widely available in seafood markets. The adductor muscles (which enable the animal to close its valves when necessary) and the retractor muscles represent the main muscular systems. The retractor muscles are a mass of strong, silky filaments by which certain bivalve molluscs, such as mussels, attach themselves to rocks and other fixed surfaces. They control the alignment and the tightness of the attachment of the mussel to the substrate. While many bivalves use their muscular foot to burrow into the sand, members of the genus Mytilus anchor themselves by using a collection of sticky threads. The mechanism by which catch is generated is not well understood but it is postulated that Molluscan catch muscle is controlled by cholinergic and serotonergic nerves. The stimulation of the former brings about an increase of the intracellular Ca^{2+} concentration, initiating phasic contraction, and the subsequent decrease of Ca^{2+} shifts it to the catch state.

Figure 79:   Fat horse mussel (*Modiolus capax*) showing bivalve structure. Mussels are common marine and freshwater bivalves. Licenced under Creative Commons Attribution-Share Alike 4.0.

Clams are also types of bivalve molluscs. The word is often applied only to those that are edible and live as animals in the sediments of the ocean floor, river or lake beds, spending most of their lives halfway buried in the sand of the seafloor or riverbeds. Hence "digging for clams". There is no reason to suppose that the electrophysiology is not basically the same as that of mussels.

Another variant bivalve mollusc is the **Razorfish** (also known as razorshell, razor clam, common razor and pod razor). They are a range of bivalve mollusc species common around the British coastline. They are an edible species of shellfish which gets its common name from their resemblance to an old-fashioned cutthroat razor. There is no reason to suppose that the electrophysiology is not basically the same as that of mussels.

Oysters are another popular edible mollusc, but the shell of the oyster is made of many fine layers, unlike the bivalve species.

Earthworms are vital in maintaining the fertility of the soil. There are 3 main types of earthworms; the compost worm, the earthworker worm and the root-dwelling worm. The natural muscle of earth-worms is an excellent actuator to drive fluids due to its membranous structure, strong force, short response time, and controllability (Tanaka, 2017) and is activated by acetylcholine-induced depolarisation (Volkov, 2001). The muscle has an electron density of about 50 mV (Volkov, 2000). The magnitude of sodium pumping was small at a normal extracellular K^+ concentration and greatly increased at higher extracellular K^+ concentrations, but these are unphysiological. The acetylcholine-sensitive receptor channel membrane complex in earthworm muscle cell differs from the acetylcholine receptor in skeletal muscle fibres and peripheral neurones of vertebrates in that it is resistant to curare (Volkov, 2001). As the mechanism otherwise seems to be similar to that of skeletal muscle, I assume that the depolarisation is caused by electron outflow.

Jellyfish

Jellyfish (Figure 80) have been in existence for at least 500 million years, and possibly 700 million years or more, making them the oldest multi-organ animal group. Jellyfish are eaten by humans in certain cultures. They are considered a delicacy in some Asian countries, where species in the Rhizostomae order are pressed and salted to remove excess water. Jellyfish and sea jellies are the informal common names given to the medusa-phase of certain gelatinous members

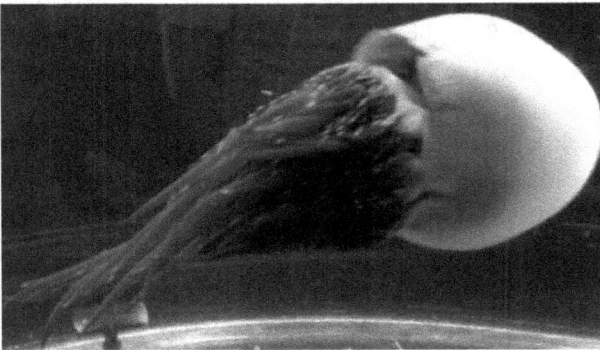

Figure 80:   Flame jellyfish. Licenced under Creative Commons Attribution-Share Alike 4.0.

of the sub-phylum Medusozoa, of the phylum Cnidaria. A green fluorescent protein used by some species to cause bioluminescence has been adapted as a fluorescent marker for genes inserted into other cells or organisms.

Jellyfish have been shown to be the most energy-efficient swimmers of all animals (Rathi, 2014). The mesoglea refers to the tissue found in jellyfish that functions as a hydrostatic skeleton. Jellyfish radially expand and contract their bell-shaped bodies to push water behind them. The pause between the contraction and expansion phases creates two vortex rings. Muscles are used for the contraction of the body, which creates the first vortex and pushes the animal forward, but the mesoglea is so elastic that the expansion is powered exclusively by relaxing the bell, which releases the energy stored from the contraction. Meanwhile, the second vortex ring starts to spin faster, sucking water into the bell and pushing against the centre of the body, giving a secondary boost forward. The mechanism, called passive energy recapture, only works in relatively small jellyfish moving at low speeds, allowing the animal to travel 30% farther on each swimming cycle. Jellyfish achieved a 48% lower cost of transport (food and oxygen intake versus energy spent in movement) than other animals in similar studies. One reason for this is that most of the gelatinous tissue of the bell is inactive, using no energy during swimming.

The complex nervous system (rhopalial nervous system, Figure 81) that controls their movements is primarily ectodermal and contains neurones immunoreactive to different antibodies (different colours in Figure 81). Their neuronal processes form multiple networks at the base of the ectoderm in the intermediate and basal segments.

The muscle of the jellyfish *Aglantha digitale* has a resting electron density index in the range of 51–82 mV, average 63 mV. The threshold of the action potential is about 20 mV electron density index. Whole-cell, patch-clamp recordings from dissociated muscle cells revealed that the inward component of the total ionic current consisted of only one calcium current. This calcium current peaked at 30 mV, and inactivated within 5 ms (Lin & Spencer, 2001). A primary function of calcium currents through this channel population is to initiate muscle

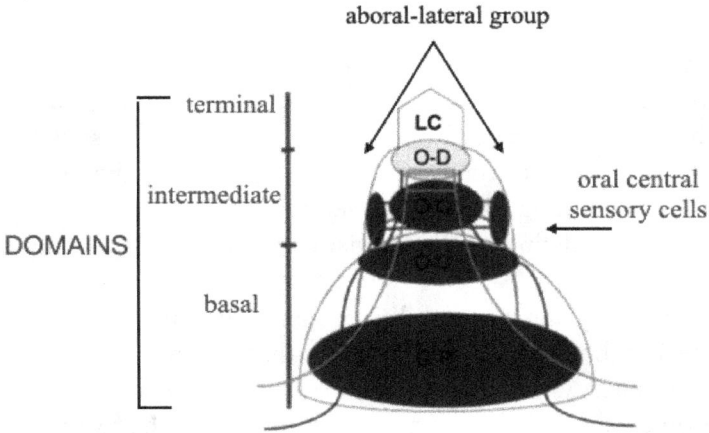

Figure 81:   Schematic diagram of rhopalium nervous system. LC: lithocyst (stato-cyst); O-D: oral-distal sensory cells. Source: Researchgate.

contraction. These authors found that the muscle feet have a far higher density of these channels than the somata (Lin & Spencer, *2001*).

Conclusion: As in every animal species I have explored, cellular electron outflow hypothetically causes depolarisation that opens a Ca^{2+} channel.

There is normal depolarisation in the absence of a sodium channel and no sodium ion inflow. As with land plants (see next chapter), depolarisation can occur in the same way for these organisms in the absence of sodium inflow. How about electron outflow?

Aequorin, used in experiments on intracellular Ca^{2+} handling (see above), was originally obtained by purification from the jellyfish *Aequorea victoria*.

Speculation: As in every animal species I have explored, muscle contraction results from Ca^{2+}-dependent activation of actomyosin and ATP. In the jellyfish example, one can postulate that this may have been the case for over 700 million years.

Deduction: Jellyfish do not have a sodium ion (Na^+) channel. Therefore Na^+ inflow cannot be a universal animal mechanism for depolarisation. Electron outflow can be.

Animals

There are innumerable other species of animals that could be described as examples of the main thesis of this work. They often possess dead parts that are no longer living cells, e.g., exoskeletons, shells, etc. But I do not want to bore the reader by describing the living cells in all these animals, ending with the same opinion.

But what is a defining characteristic that distinguishes animal from plant?

I suggest that one is the ability of many animal species (not anaerobic ones) to breathe oxygen and use that oxygen to obtain energy from food, energy that can be used to fuel locomotion.

However, the quantum structure of oxygen in the form O_2 reveals that in its most stable ground state, O_2 exists in air and is a potentially toxic, mutagenic free radical gas, because it has two unpaired electrons with parallel spins in opposing orbitals (Bailey, 2019).

I quote Bailey (2019), "Thankfully for us, this configuration means that O_2 is 'spin restricted', forcing it to accept electrons one at a time, with the sequential formation of free radicals and reactive oxygen species during its reduction to water in the mitochondria......" Indeed, if it was not for this thermodynamic quirk of fate, we would combust spontaneously in room air.

CHAPTER 32

Electricity passing through flesh

Up to now, I have been pursuing the story of electron flow depolarisations in various animal tissues and in various animal species, but the original idea was that if one can record electricity on the body surface, the animal must be generating electricity. I presented as an example the human electrocardiograph (Figure 2).

From cardiac action potentials to ECG

I will remain with this example because I know something about electrocardiographs (ECGs). At the Hammersmith Hospital in 1966, one of my duties every morning was to write a report on every ECG recorded in the hospital inpatients and outpatients. My reports were checked and criticised by a consultant cardiologist. In those days, the usual 12 standard leads were recorded separately on long narrow strips of paper, not on one sheet like the example (Figure 2). I continued with looking at ECGs for the rest of my career.

For this process to make any sense one must first describe the anatomy from which these electrical signals come within the heart (Figure 82).

But to correlate all this to what we might find on the body surface, the timing of the events is crucial (Figure 83).

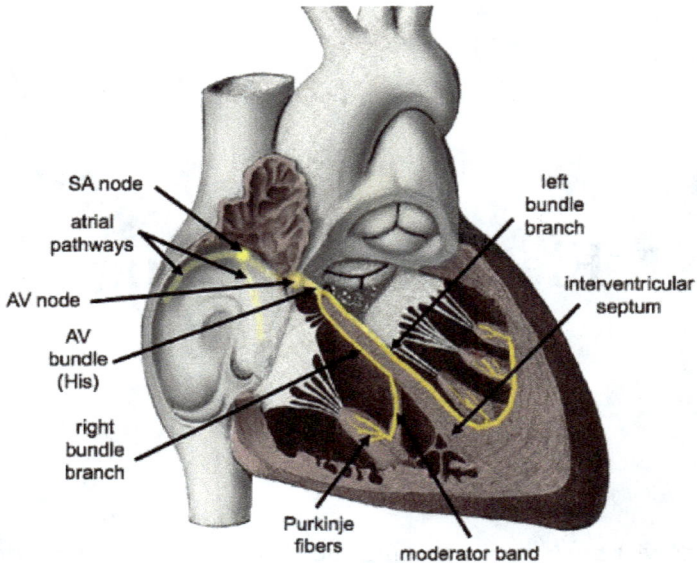

Figure 82:   The route through the heart of the electrical impulse (i.e., action potential) generated in the sinus node, thenceforth conducted through atria tissue to the atrio-ventricular node, His bundle and bundle branches and then to the ventricles (brown). Licenced under Creative Commons Attribution-Share Alike 4.0.

The most important feature of the timing is the delay in transmission of the electrical impulse through the atrio-ventricular (AV) node and bundle of His. This ensures that there is a delay of nearly 100 msec between atrial and ventricular activation and therefore a similar delay in contraction, so that atrial contraction completes the filling of the relaxed ventricles before the latter contract. According to Mezzano (*2014*) the role of cardiomyocyte cell junctions in the cardiac conduction system remains unclear. It seems to me that if outer insulation is high and accompanied by increased electrical impedance of the gap junctions, the speed of spread of action potential through the AV node and His bundle is predicted to be slow. Failure of the delay, e.g., by an accessary pathway bypassing the AV insulating disc, causes arrhythmias. (Pathophysiology is not in my remit for this work; whole books are available describing cardiac rhythm disturbances).

I see no useful purpose in this context of inverting all the action potentials as I did, for instance, in Figures 11 and 12, and no useful

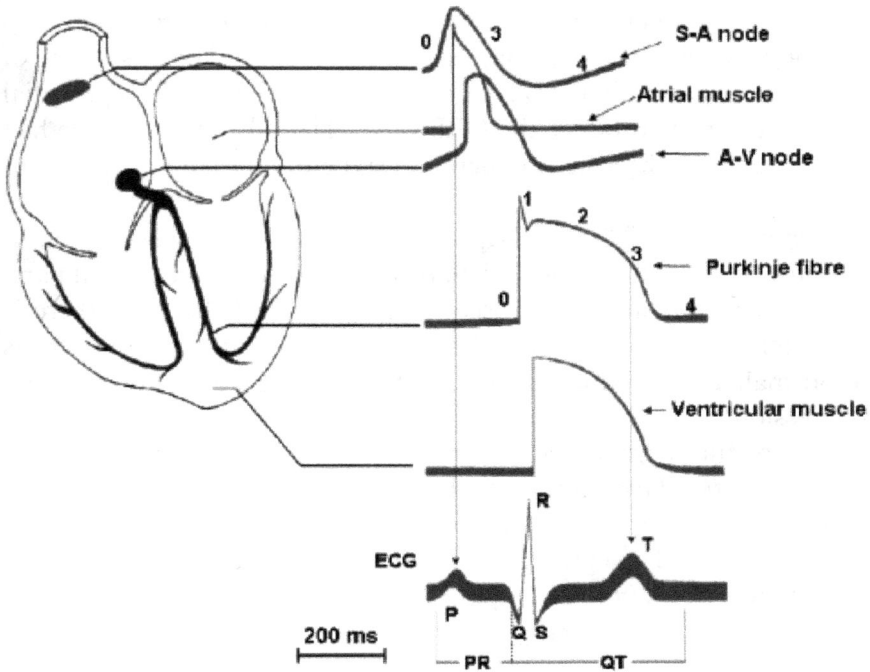

Figure 83:   The changes in action potential configuration and timing as the electrical signal passes through the heart. All action potentials are displayed in conventional polarity, i.e., I have not inverted them to emphasise that they are reductions in electron density. The diagram adopted the conventional idea of ions flows rather than electron flows and these were numbered. I request the reader to ignore the numbers. S-N = sino-atrial node; A-V node = atrioventricular node. Source: Researchgate.

purpose in inverting all the signals in the ECG (Figure 2). That would not affect any conclusion here, and there is no chance that such an action would be generally adopted; it would not affect ECG inter- pretation. The other important features of the timing are the delay in depolarisation of the epicardium with respect to the endocardium, and the delay in repolarisation of the endocardium with respect to the epicardium. The corresponding ECG waves are caused by differ- ences in timings. The P wave indicates a difference in timing of atrial depolarisation, but, because the repolarisation of the atria is slow (compared with speed of depolarisation), no difference in timing is great enough to cause a potential difference that is transmitted to the ECG. However, in the case of the QRS wave of the ECG, the cause is due to the earlier depolarisation of the endocardium for a few milliseconds while the epicardium remains polarised. There is a sort of subtraction (a simplification) of endocardial voltage from epicardial voltage to partly produce the QRS wave. The other factor is the fact that the apex of the ventricles depolarises before the base, as is evident in Figure 82.

Between the QRS wave (depolarisation) and the T wave (repolarisation), the ECG does not deviate from zero voltage because all parts of the ventricles are depolarised at this time and therefore there is no potential difference. Deviation of this "ST segment" from zero is abnormal. These abnormalities may be called "ST elevation" or "ST depression".

I sometimes get asked, "If the QRS represents depolarisation, should not repolarisation be represented by a negative wave, but the T wave is positive?" It is easy to see from Figure 83 that this is a misunderstanding. The epicardium repolarises first so that, once again, subtraction of its voltage (repolarised) from the endocardial action potential (still depolarised) gives a positive deflection.

Passage of some electricity from the heart to the body surface

According to conventional theory, positively charged potassium ions leave the heart cells, but I have never heard anyone suggest that these flow on to the body surface. In the electron theory, once electrons leave the heart cells to produce depolarisation, they, or the electromagnetic wave, meet a very different electrical impedance — the tissue of the body. Passage of this electricity through various impedances will involve electron losses accompanied by heat. Only fairly large electricity sources will reach from the heart to the body surface, much attenuated by heat losses (as with all electric current flows). The wave will, in any case, travel in all directions unless all potential pathways have the same impedance. They do not. One has to consider the variable anatomy, e.g., bone, cartilage, muscle, etc., each of which will present different impedances. This fact has led to the subject of vectorcardiology. The vectorcardiogram represents heart vector movement in three orthogonal dimensions and the changes with time. The reader will find the complexity of this approach very difficult.

Frontal plane (2D) vector analysis

For the average reader (and myself), vector loop analysis is difficult, but I nevertheless use the vector system in reading ECGs, particularly the standard leads I, II, III, AVR, AVL and AVF (Figure 84). Imagine laying this triangle on a person's torso.

A vector has magnitude and direction. If one looks at the normal ECG, one observes that most QRS deflections are monophasic. That means that during the time that the QRS lasts, the directions of the vectors are roughly the same and one can look at the area under the curve and say that that is the direction of the mean vector for that lead. Then one can observe that of leads I, II and III in Figure 2, the largest vector is in lead II. So the main QRS vector in this case is near

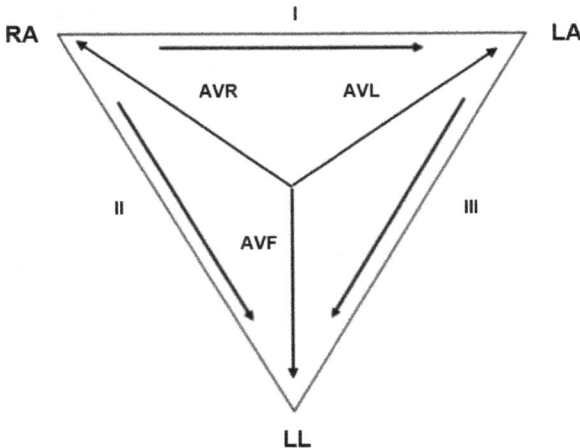

Figure 84: Simple triangular 2D representation of standard leads to indicate vector directions. RA = right arm lead; LA = left arm lead; LL = left leg lead (the right leg lead is a relative "earth lead"). Lead I indicates the LA voltage minus the RA voltage. Lead II is the left leg voltage minus the right arm lead voltage. Lead III is the left leg voltage minus the left arm voltage. The central point voltage on the triangle is derived from the first 3 standard leads and is subtracted from the RA voltage to obtain AVR; from the LA voltage to obtain the voltage of AVL and; from the LL lead to obtain the voltage of AVF. From my teaching notes.

the direction of lead II which is downwards and to the left. In order to simplify this deduction, I present Figure 85.

If one now looks at leads AVR, AVL and AVF on the ECG (Figure 2), one observes that the largest mean vector is a negative one in lead AVR, so it is going in the opposite direction to lead AVR, indicated by minus AVR (−AVR) in Figure 85. Also lead AVL is small in amplitude and, although biphasic, the mean is clearly close to zero. That means that the maximal mean vector is at right angles (90°) to lead AVL (Figure 85), confirming that it is the direction of lead II.

The direction for the possible mean maximal amplitude QRS vector is indicated in Figure 85. If I find that mean maximal QRS vector is outside direction III, I report "right axis deviation" (abnormal). If I find that mean maximal QRS vector is higher than lead AVL, I report "left axis deviation" (also abnormal). If the QRS is biphasic, one must work out the directions of the early and late parts and if there is a wide angle between them, this is abnormal. If a T wave does not deflect in the same direction in a standard lead, one must work out the directions of the QRS and T vectors and if there is a wide

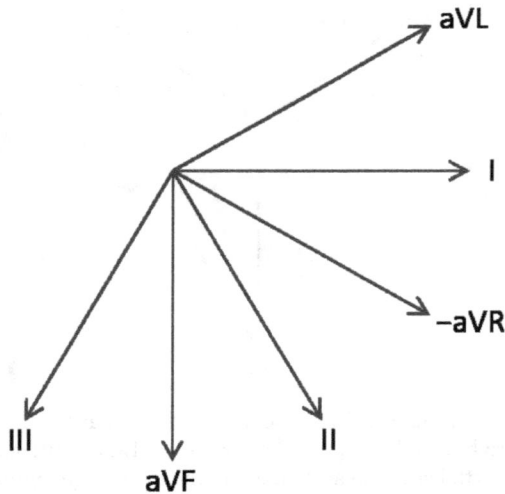

Figure 85: Direction of the maximum mean vector in the frontal plane if it falls on a standard lead. Most normal people have a mean QRS vector in directions between leads I and AVF (ignore −AVR which is in the opposite direction to AVR). From my teaching notes.

angle between them, it is abnormal. This abnormality is often called "T wave inversion".

Anterior to posterior vector analysis

The approach to this is more difficult because one has to rely on unipolar chest lead voltages (Figure 86).

Inspection of the voltages recorded from chest leads V_1 and V_2 show maximum mean deflections that are negative, because the left ventricle which dominates the electrical events is behind these lead positions. By contrast leads V_4, V_5 and V_6 show positive mean deflection indicating a lateral direction.

Electricity passes through the body with attenuation due to impedance and consumption, with conversion to heat. The voltages recorded at the surface are therefore of much less magnitude than those at the organs' source.

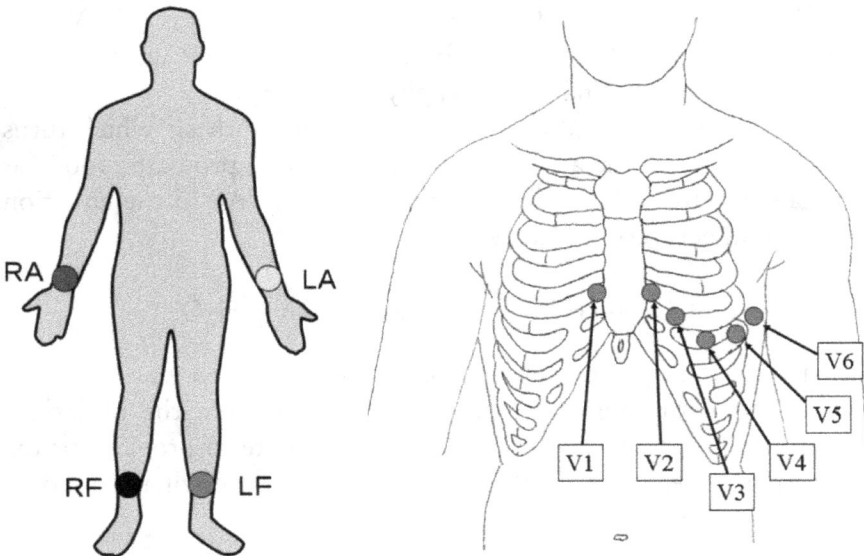

Figure 86:   The positioning of the limb leads (left) and chest leads V_1–V_6 (right). Licenced under Creative Commons Attribution-Share Alike 3.0.

CHAPTER 33

Plants

Land plants

The fact that electron flows from the living cell considered thus far are universal in the animal kingdom, I wondered whether there is anything similar in plants. The answer is clearly "yes". Ulirich & Novacky (*1991*) studied the electrical properties of leaves, roots, and single root cap cells of the common oat.

The electron density of the living cells of this plant is 80 mV. The depolarisations studied by Ulirich & Novacky (*1991*) were induced by a specific toxin. What happens physiologically?

A distinguishing feature of plants is photosynthesis which turns carbon dioxide in the air into organic molecules, providing food for animals. Photosynthesis depends on electrons. "A" in the equation below is an acceptor molecule.

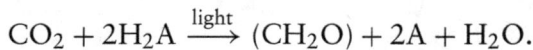

$$CO_2 + 2H_2A \xrightarrow{\text{light}} (CH_2O) + 2A + H_2O.$$

Most plants are green (Figure 87); speculation as to why this is so ranges from random chance to the possibility that the radiation-absorbing properties of chlorophyll are adequate to provide for the light energy needs of Earth's plants. The chlorophyll is found in chloroplasts (Figure 88).

Figure 87:   Red oak's young leaves. Licenced under Creative Commons Attribution-Share Alike 4.0.

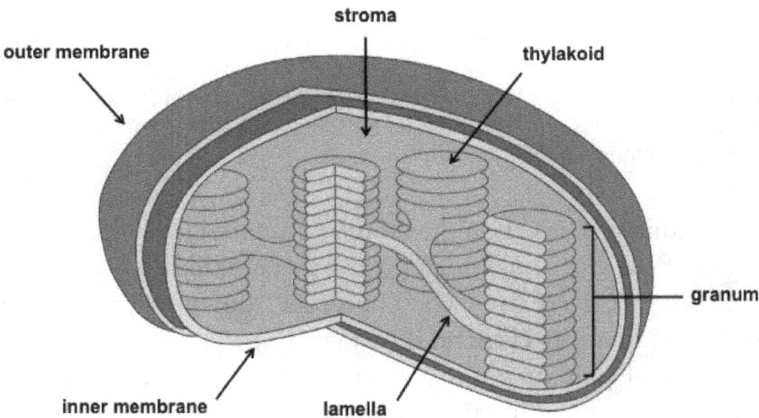

Figure 88:   Cutaway impression of a chloroplast. Licenced under Creative Commons Attribution-Share Alike 4.0.

The pathway of electrons

The general features of a widely accepted mechanism for photoelectron transfer, in which two light reactions (light reaction I and light reaction II) occur during the transfer of electrons from water to carbon dioxide, were proposed by Robert Hill and Fay Bendall in 1960 after extensive research (Hill, *1937*, *1939*; Hill & Scarisbrick, *1940*).

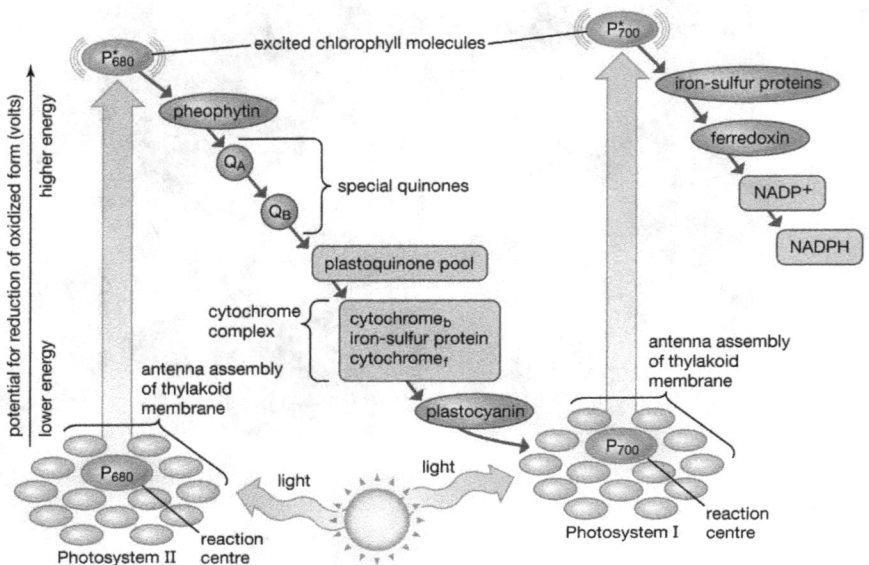

Figure 89: Flow of electrons during the light reaction stage of photosynthesis. From P_{680} (bottom of left thick arrow), photosystem II extends diagonally down to photosystem I (bottom of right thick arrow) to initiate it. P_{700} initiates photosystem I, commencing the electron pathway from P_{700} diagonally down to NADPH. Source: Wikimedia Commons, public domain.

This mechanism is based on the relative potential (in volts) of various cofactors of the electron-transfer chain to be oxidised or reduced.

In Figure 89, there are two "staircases" down to the right. Each "step down" involves **electron flow** (arrows).

The special pair of photosystem I is called P_{700}, while the special pair of photosystem II is called P_{680}. When light is absorbed by one of the pigments in photosystem II, energy is passed inward from pigment to pigment until it reaches the reaction centre to trigger photosystem I. The end product is nicotinamide adenine dinucleotide phosphate, abbreviated $NADP^+$. $NADP^+$ is a coenzyme that functions as a universal electron carrier, accepting electrons and hydrogen atoms to form NADPH. $NADP^+$ is created in anabolic reactions, or reactions that build large molecules, e.g., organic food molecules, from small molecules.

Organic compounds are formed by using carbon dioxide as a carbon source. The rates of these reactions can be increased somewhat by increasing the carbon dioxide concentration. Since the middle of the 19th century, the level of carbon dioxide in the atmosphere has been rising because of the extensive combustion of fossil fuels and deforestation. The atmospheric level of carbon dioxide climbed from about 0.028% in 1860 to 0.032% in 1958 (when improved measurements began) and to 0.041% in 2020. This increase in carbon dioxide directly increases plant photosynthesis up to a point, but the size of the increase depends on the species and physiological condition of the plant.

Restricted water availability limits photosynthesis and plant growth. Large amounts of water are transpired from the leaves; that is, water evaporates from the leaves to the atmosphere via the stomata, small openings through the leaf epidermis, or outer skin. They permit the entry of carbon dioxide but inevitably also allow the exit of water vapour. The stomata open and close according to the physiological needs of the leaf. In hot and arid climates the stomata may close to conserve water, but this closure limits the entry of carbon dioxide and hence the rate of photosynthesis. The decreased transpiration means there is less cooling of the leaves and hence leaf temperatures rise.

Plants require more than light, water and carbon dioxide, but also minerals that they must obtain from the soil. (After he retired, I visited the home of Mullard, the great thermionic valve manufacturer, where he was experimenting with soil-less plant growth.) It is not that easy to get the soil (or soil-less) composition optimal, but constant spraying with chemical fertiliser is to be deprecated as some of them are toxic and are entering the food chain.

But plants do not just photosynthesise; they have work to do such as growth and therefore show tissue respiration. Plants depolarise and show action potentials. This is known not to be due to an inward Na^+ current owing to the lack of a suitable Na^+ gradient and to the fact that Na^+ ions are toxic to most land plant cells. I suggest that both plant and animal cells depolarise as the result of an electron outflow eliciting action potentials by opening Ca channels. The conventional explanation for this depolarisation seems to be an outflow of chloride ions (Cl^-), but then how is Cl^- balance achieved? With depolarisation by electron outflow, there is no such problem. Why do biologists always insist on postulating ionic flow rather than electron flow? As plants also have mitochondria, I postulate the same mechanism that I did for animals, namely depolarisation by electron outflow from the cell, and repolarisation by electron flow from mitochondria to the cytoplasm.

Vodeneev (2015) describes variable potentials (VP) rather than depolarisation and repolarisation. These authors think that mechanisms of VP generation and propagation are uncertain and still under investigation. They also think that it is probable that VP is a local electrical response induced by propagation of a hydraulic wave and/or chemical agent. Both hypotheses are based on numerous experimental results but they predict that VP velocities are not in good accordance with speed of VP propagation (!) and that a combination of hydraulic and chemical signals is the probable mechanism of VP propagation. VP generation is traditionally connected with transient H^+-ATPase inactivation, but action potential spikes are also connected with passive ions fluxes. These authors also propose that Ca^{2+} influx is a probable mechanism that triggers H^+-ATPase inactivation (Vodeneev, 2015). This seems to me an unnecessarily complicated description; I would more simply state that some partial depolarisations take place that do not reach the threshold for triggering action potentials and openings of the Ca^{2+} channel. An extended discussion of the older conventional ideas on this subject is presented by Higinbotham (1973).

Plant physiologists focus on plant polarity which is essential for processes such as intercellular communication, cell division, cell morphogenesis and differentiation. Polarised cells are involved in the patterning of plant tissues and organs, and therefore contribute to the overall shape of the plant.

An intriguing example of plant intercommunication occurs in tomato plants. Tomato plants respond to herbivore attack by emitting volatile organic compounds, which are released into the surrounding atmosphere. Zebelo (2012) tested the ability of tomato receiver plants to detect tomato donor volatiles by analysing early responses, including electron density variations and cytosolic Ca^{2+} fluxes. Receiver tomato plants responded within seconds with a strong depolarisation, which was only partly recovered by blowing receiver plants with clean air. Green leaf volatiles were found to exert the stronger depolarisation, when compared to α-pinene and β-caryophyllene. Depolarisation was found to increase with increasing green leaf volatile concentration and were also found to induce a strong Ca^{2+} increase, particularly when (Z)-3-hexenyl acetate was tested both in solution and with a gas (Zebelo, 2012). On the other hand, α-pinene and β-caryophyllene, which also induced a significant depolarisation with respect to controls, did not exert any significant effect on Ca^{2+} homeostasis. Plant perception of volatile cues (especially green leaf volatiles) from the surrounding environment is mediated by early events, occurring within seconds and involving the alteration of electron density and Ca^{2+} flux. These observations are consistent with the idea that green leaf volatiles reduce the cell impedance of tomato receiver plants. An example of a herbivore that stimulates this type of reaction is *Tuta absoluta* larvae (Anastasaki, 2018).

I have searched for data on the threshold for depolarisation. Brault (2004) studied a suspension of cells from Arabidopsis, a small flowering plant related to cabbage and mustard. Depolarisation is triggered by abscisic acid, a plant hormone that functions in many plant developmental processes, including seed and bud dormancy, the control of organ size and stomatal closure. I would have preferred a study of measurements of electron density in the cells under normal conditions, but they used voltage clamps. They reported that abscisic acid-induced depolarisation was the result of two different processes: activation of anion channels and inhibition of H^+-ATPases. These two processes are independent because impairing one did not suppress the depolarisation. Both processes are, however, dependent on the Ca^{2+} increase induced in the cytoplasm since an increase in Ca^{2+} enhanced anion channels and impaired H^+-ATPases. I repeat my objection to the assumption of anion outflow to cause depolarisation because it creates ion imbalance, as does cation inflow in animal cells. This problem does not occur if one assumes electron outflow. However, the study appears to provide evidence of a decrease in proton pumping during depolarisation (Brault, 2004).

A more physiologically interesting intervention in plants is the response to darkness, in view of the dependence on light for photosynthesis. Kupisz (2017) studied thalli (singular thallus, composed of filaments or plates of cells and ranging in size from a unicellular structure to a complex treelike form, with a simple structure that lacks specialised tissues typical of higher plants) of *Marchantia polymorpha*, sometimes known as the common liverwort or umbrella liverwort, a large liverwort with a wide distribution around the world. It is variable in appearance and has several subspecies. It is dioicous, having separate male and female plants (i.e., dioicous plants are those that have gametophytes which produce only sperm or eggs but never both). These authors found that darkening caused persistent depolarisation of the electron density and generation of short-lasting potential changes that were not uniform among different thalli. In some plants (18%), the changes evoked by darkening were typical action potentials (APs_{dark}), whereas in 69% of the plants, the changes had a form of action potential-like responses ($APs_{dark-like}$) consisting of a transient depolarisation followed by a plateau phase, whose magnitude and duration were inconsistent. The illumination of the *M. polymorpha* always evoked action potentials (APs_{light}) if the thallus was illuminated after 30-min darkening. To analyse the involvement of H^+-ATPase in formation of the illumination/darkening-induced electrical responses in *M. polymorpha*, proton pump regulators were used. One proton pump inhibitor significantly diminished the resting electron density and inhibited dark-induced APs_{dark} and/or $APs_{dark-like}$ responses and illumination-induced APs_{light}. After application of another proton pump inhibitor, there was strong depolarisation and no response to light/dark. An activator of the proton pump strongly hyperpolarised electron density and blocked dark-induced APs_{dark}/$APs_{dark-like}$ responses and illumination-induced APs_{light}.

These effects are illustrated in Figure 90. My interpretation is that in the case of the so-called dark-like action potentials, there is an initial fall in electron density (after a minute transient increase). The electron density fall reaches a threshold at the value labelled "initial depolarisation" (Figure 90), which triggers an action potential with Ca^{2+} inflow, i.e., a spike of loss of electron density, immediately followed by a rapid phase of repolarisation. There is then a plateau of persistent low electron density that is followed by complete repolarisation (electron flow from mitochondria to cytoplasmic resting electron

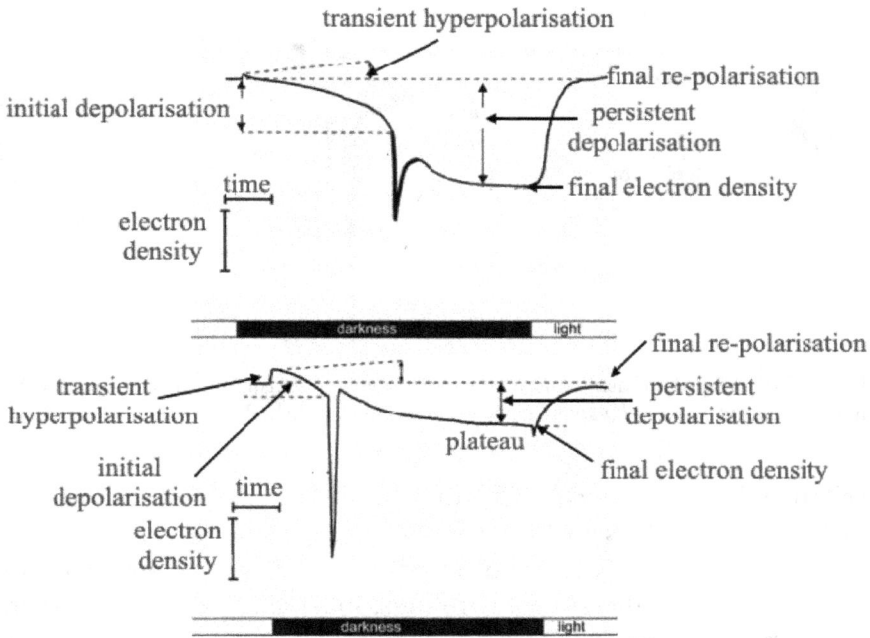

Figure 90:   First presentation of two types of electron flows (above and below) during activation of the liver wart *Marchantia polymorpha*. Not previously published.

density level). In the case of dark action potentials (lower example in Figure 90), there are the same changes except that the transient hyperpolarisation (cytoplasmic electron gain) is greater, the initial repolarisation is greater and the persistent depolarisation is less. In both cases repolarisation rapidly follows the resumption of light after the darkness period.

Sea plants

The unicellular marine diatom (a major group of algae, specifically microalgae, found in the oceans, waterways and soils of the world) *Odontella sinensis* can generate a fast action potential (Figure 91) that has remarkably similar biophysical and pharmacological properties to invertebrate and vertebrate cardiac and skeletal muscle cells. This has induced Taylor (2009) to assume that depolarisation is carried by sodium ions (Na^+). This seems crazy to me. Why should land plant

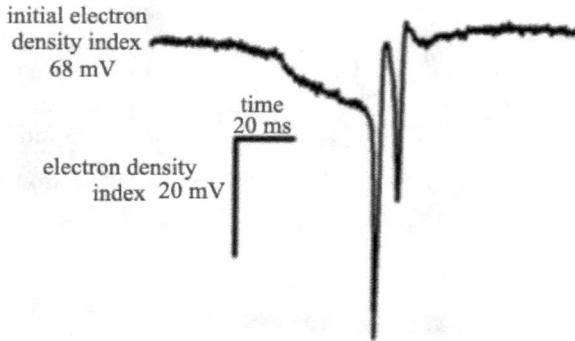

Figure 91:   First demonstration of changes in electron density upon activation of the marine diatom *Odontella sinensis*. Not previously published.

cells depolarise by chloride (Cl^-) outflow and marine plants depolarise by Na^+ inflow? Does it not make more sense to assume that they both depolarise by the same mechanism, namely electron outflow? How do the ion flow hypotheses deal with ionic imbalance? There is no such problem with the electron flow hypothesis.

CHAPTER 34

Fungi

Having considered the electrophysiology of animals and then plants I have missed out fungi. Fungi are not plants. Living things are organised for study into large, basic groups called kingdoms. Fungi were listed in the plant kingdom for many years. Then scientists learned that fungi show a closer relation to animals, and are unique and separate life forms.

The opportunistic fungal pathogen, *Candida albicans*, regulates the electron density in response to environmental conditions, as well as the physiological state of the cell. Suchodolski & Krasowske (*2019*) used Di-4-ANEPPS fluorescence to explore electric potential difference across the cell membrane of this fungus. The electron density of this fungus is approximately 120 mV, whereas that of the non-pathogenic yeast *Saccharomyces cerevisiae* is only about 70 mV. These authors demonstrated growth-dependent depolarisation (electron outflow) during the growth curve of *C. albicans* CAF2-1, strain grown to 8 h and 14 h (Suchodolski & Krasowske, *2019*).

These examples, if showing general biological properties of fungi, suggest that the electrical events are similar in principle to those of animals and plants. The electron densities tend to be higher than those in animal and plants and electron flows seem to initiate functions.

CHAPTER 35

"Primitive" organisms

Humans have often speculated on the subject of "How did life begin?" This cannot be answered as there was no human observer when this happened. Then we started extrapolating backwards through study of species that were supposed to be represent early species, e.g., single cell organisms.

Amoebae

Many students will be familiar with these organisms. Amoebae are single-celled animals that catch food and move about by extending fingerlike projections of protoplasm. Phagocytosis is a process by which certain living cells called phagocytes ingest or engulf other cells or particles. A phagocyte may be a free-living one-celled organism, such as an amoeba, or one of the body cells, such as a white blood cell.

Amoebae are either free-living in damp environments, or parasitic. As with the other cells described, their DNA is packaged into a central cellular compartment called the nucleus. Amoebae contain no mitochondria and generate ATP exclusively via anaerobic means. Nevertheless they display intracellular electron density. It appears that this occurs in response to an externally applied direct current (De la Fuente, *2019*). Non-physiological interventions have been studied. Potassium and sodium ions decrease the electron density by 40 and 35 mV, respectively, per tenfold increase of external concentration.

A transient biphasic shift of electron density with the spontaneous emptying of the contractile vacuole has been recorded. Anionic compounds do not depolarise the membrane. It is suggested that the permeability of the cell membrane is largely governed by negatively charged groups of the mucopolysaccharide layer of the membrane (Josefsson, *1966*).

Slime moulds

Slime moulds are several kinds of unrelated eukaryotic organisms that can live freely as single cells, but can aggregate together to form multicellular reproductive structures. Slime moulds were formerly classified as fungi but are no longer considered part of that kingdom. Vallverdu (*2018*) thought that the slime mould *Physarum polycephalum* might show cognition, a primitive type of consciousness.

This *Physarum polycephalum* is an acellular slime mould or myxomycete (a class of slime moulds that contains 5 orders, 14 families, 62 genera, and 888 species) with diverse cellular forms and broad geographic distribution. Microplasmodia of *Physarum polycephalum* have been investigated by conventional electrophysiological techniques. In standard medium the electron density is around 100 mV, calculated from the measurements of Fingerle & Gradmann (*1982*). This is insensitive to light and changes of the Na^+/K^+ ratio in the medium. Without bivalent cations in the medium the electron density drops to about 0 mV, as in other living cells, i.e., bivalent cations in the external medium are required. By analogy with other living cells, the physiological cation involved is probably Ca^{2+}. Addition of glucose or sucrose (but not mannitol or sorbitol) causes rapid depolarisation, which partially recovers over the next few minutes. Half-maximal peak depolarisation (25 mV with glucose) was achieved with 1 mM of the sugar. There is an electrogenic H^+ extrusion (proton) pump with a stoichiometry of 2 H^+ per metabolic energy equivalent; the de-protonated form of the pump seems to be negatively charged (electron excess).

Plankton

Plankton are the diverse collection of organisms found in water (or air) that are unable to propel themselves against a current (or wind). The individual organisms constituting plankton are called plankters. They form the beginning of the food chain for many creatures, including fish and whales, from which they enter the food chain for land animals that catch or hunt them. Plankton includes plants and animals that float along at the mercy of the sea's tides and currents. Their name comes from Greek, meaning "drifter" or "wanderer". There are two types of plankton: tiny plants called phytoplankton, and weak-swimming animals called zooplankton. Phytoplankton, as described for land plants above, use sunlight, carbon dioxide (CO_2) and water in photosynthesis, involving an electron chain (Figure 89) to produce organic compounds which they use for food and to make their cells. Zooplankton are the animal component of the planktonic community ("zoo" comes from the Greek word for animal). They are heterotrophic (other-feeding), meaning they cannot produce their own food and must consume instead other plants or animals as food. In particular, this means they eat phytoplankton. Figure 92 illustrates the plankton food cycle.

Gradman & Boyd (1995) studied the phyloplankton *Coscinodiscus radiatus* using conventional intracellular glass micro-electrodes, and recorded a membrane potential in a standard electrolyte environment of around -85 mV, implying an electron density index of 85 mV. Membrane voltages much more negative, e.g., up to -140 mV (electron density index, i.e., root mean square, of 140 mV) have been recorded. Within the voltage range, light on and off (microscope illumination) caused weak hyperpolarisations and depolarisations by about 2 mV with a time constant of about 10 seconds. Also within the voltage range, spontaneous oscillations could be observed with a frequency of about 0.03 Hz and irregular amplitudes up to 30 mV. Electro-coupling worked out for guard cells with their subtle osmoregulatory system may be equivalent mechanisms in planktonic diatoms for adjustment of buoyancy by appropriate uptake and release of ions.

Feeding modes in the freshwater zooplankton include phagocytosis, filter feeding, and individual particle selection. Each of these

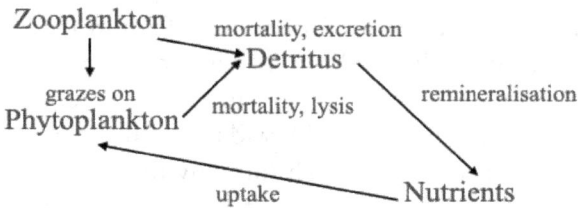

Figure 92: The simplest example of plankton nutrient recycling. There are many more complicated examples. Fundamentally the main food source is photosynthesis in phytoplankton, the start of the food chain for many organisms. Not previously published.

modes has different relationships between the size of the predator and the size of the prey. Zooplankton are a trophic bridge between the remainder of the ecosystem. Herbivorous zooplankton in lakes sometimes greatly reduce the abundance of planktonic algae.

Species of zooplankton are not dispersed uniformly or randomly within a region of the ocean. As with phytoplankton, 'patches' of zooplankton species exist throughout the ocean. Some species of zooplankton are restricted by salinity and temperature gradients, while other species can withstand wide temperature and salinity gradients. Zooplankton patchiness can also be influenced by biological factors, including breeding, predation, concentration of phytoplankton, and vertical migration. The greatest factor is mixing of the water column (upwelling and downwelling along the coast and in the open ocean) that affects nutrient availability and, in turn, phytoplankton production.

Nowadays, a great danger to the food chain is the ingestion by zooplankton of microplastics (Cole, 2020).

Implication: **Animals that depend on plants (directly or indirectly through the food chain) therefore depend on the plant's photosynthesis by light photons and electrons. Photosynthesis depends on water supply. Therefore, life on Earth depends on water supply.**

Hydra

Hydra is a genus of simple freshwater animals belonging to the phylum Cnidaria, a phylum containing over 9,000 species found only in aquatic and mostly marine environments. Phylum Cnidaria includes animals that exhibit radial or biradial symmetry and are diploblastic, meaning that they develop from two embryonic layers, ectoderm and endoderm. Nearly all (about 99%) cnidarians are marine species. The family includes corals and jellyfish (already discussed). In the apparent absence of data, one might assume electron densities similar to jellyfish (about 60 mV).

Lamprey

Lampreys are thought to be an ancient extant lineage of jawless fish of the order Petromyzontiformes. Ammocoetes (the filter-feeding larvae of modern lampreys) have long influenced hypotheses of vertebrate ancestry. The life history of modern lampreys, which developed from a superficially amphioxus-like ammocoete to a specialised predatory adult, appears to recapitulate widely accepted scenarios of vertebrate origin. However, according to Miyashita (*2021*), no direct evidence has validated the evolutionary antiquity of ammocoetes, and their status as models of primitive vertebrate anatomy is uncertain. Miyashita (*2021*) reported that larvae of all four genera lack the defining traits of ammocoetes. They instead display features that are otherwise unique to adult modern lampreys, including prominent eyes, a cusped feeding apparatus, and posteriorly united branchial baskets. These authors are of the opinion that their results indicate that ammocoetes are specialisations of modern lamprey life history rather than relics of vertebrate ancestry. The resting calculated electron density is about 75 mV and depolarisation causes Ca^{2+} entry (Wang, *2013*).

CHAPTER 36

Bacteria

Bacteria are small single-celled organisms, found almost everywhere on Earth and are vital to the planet's ecosystems. Some species can live under extreme conditions of temperature and pressure. Although bacteria do not contain mitochondria, they do exhibit oxidative phosphorylation (Meyer & Jones, 1973), which separates electrons and protons. Combined with proton pumping (Friedrich, 1995) this results in higher electron density in bacteria than in mammalian muscle (more negative intracellular voltage in the conventional system). Thus, bacteria generate electron density and ATP production by similar mechanisms to the mitochondria of "higher" organisms.

While a number of diseases are caused by bacterial infection, many have beneficial effects on all other life forms. Some bacteria are anaerobic, i.e., do not use oxygen. They are similar to **Archaea**, which are single-celled microorganisms with structure similar to bacteria. They are evolutionarily distinct from bacteria and eukaryotes and form the third domain of life. They live in environments low in oxygen (e.g., water, soil).

Felle (1980) studied growth in giant cells of *Escherichia coli* (Figure 93) stimulated by 6-amidinopenicillanid acid. Measurements by the steady-state distribution of [^{3}H] tetraphenylphosphonium agreed closely with those made with intracellular micro-electrodes. Calculation from these measurements indicate an electron density index of 85 mV at pH 5.0 and of 142 mV at pH 8.0 (average 22 mV per pH unit between pH 5.0 and 7.0). *Escherichia coli* also depolarise,

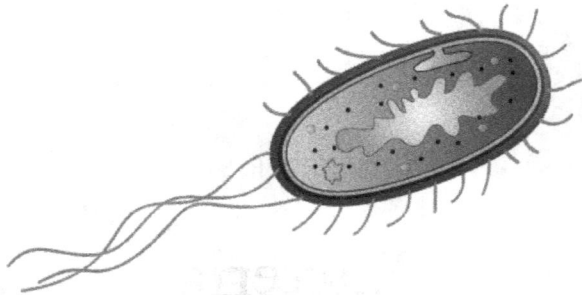

Figure 93:   Diagram of the bacterium *Escherichia coli*. Images are from Togo picture gallery maintained by Database Center for Life Science (DBCLS). Licenced under Creative Commons Attribution-Share Alike 3.0.

indicating electron loss, during growth. Four regions of the exponential growth stage investigations indicated that the electron density of 220 mV in the early exponential phase depolarises to 140 mV in the late exponential phase (Bot & Prodan, *2010*).

Electron density changes regulate a wide range of bacterial physiology and behaviours, for example, pH homeostasis, membrane transport, motility, antibiotic resistance, cell division, electrical communication, and environmental sensing. Benarroch & Asally (*2020*) reviewed the physiological roles of bacterial membrane potential as a source of free energy and as a means of information signalling and processing. Bacterial electron outflow can cause a "persister" state involving more resistance to antibiotics.

Antibiotic resistance in pathogenic bacteria, e.g., *Staphylococous aureus* (MRSA), is a large problem encountered by doctors treating infections. At pH 5.0, the electron density of *Staphylococcus aureus* exhibits a minimum of 85–90 mV. With increased pH, a maximum of 100 mV is reached.

Antibiotic resistance is probably mainly due to changes in the bacterial genome. Vestergaard (*2018*) explored the Nebraska Transposon Mutant Library of 1920 single-gene inactivations. *S. aureus* strain JE2 was screened for reduced susceptibility to the antibiotic, gentamicin. Nine mutants were shown to display between 2-fold and 16-fold reduced susceptibility to this antibiotic. All of the identified genes were associated with the electron transport chain and energy metabolism.

Four mutant strains conferred the largest decrease in gentamicin susceptibility and three exhibited a small colony variant phenotype,
whereas the remaining mutants displayed colony morphology similar
to the wild type. All of the mutants, except hemX, displayed reduced
electron density. The results demonstrate that *S. aureus* possesses multiple genes that, upon inactivation by mutagenesis, reduce the electron
density, thus reducing the lethal activity of gentamicin.

Microbiome

The human body is full of bacteria, and is estimated to contain more
bacterial cells than human cells. The so-called microbiome is the
genetic material of all the microbes — bacteria, fungi, protozoa and
viruses — that live on and inside the human body. The number of
genes in all the microbes in one person's microbiome is 200 times the
number of genes in the human genome. The microbiome may weigh
as much as five pounds (heavier than the brain!). While bacteria are
the biggest players, we also host single-celled organisms known as
archaea, as well as fungi, viruses and other microbes, including viruses
that attack bacteria. Together these are dubbed the human microbiota. The human body's microbiome is all the genes your microbiota
contains; however, colloquially, the two terms are often used inter-
changeably.

A plethora of conditions, from obesity to anxiety, appear to be
linked to these microbes. Spector (2020), in particular, pleads for
respect for the integrity of the microbiome, which can be disturbed by
antibiotics and highly processed foods. All junk foods are highly processed and should be avoided. The changes induced in the microbiome
by bad diet and drugs can last for months, causing ill health.

Archaea

Archaeal cells (Figure 94) have unique properties separating them
from Bacteria and Eukarya. Archaea are further divided into multiple recognised phyla. Classification is difficult because most have
not been isolated in a laboratory and have been detected only by
their gene sequences in environmental samples. Several metabolic
pathways are more closely related to those of eukaryotes, notably for

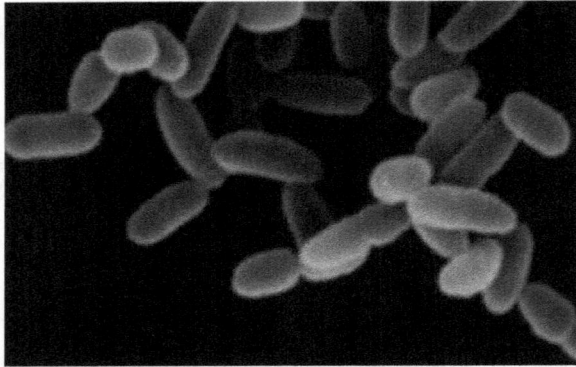

Figure 94:   Archaea. The White Ave started as an Archaea, a single-celled microorganism; each cell is about 5 μm long. Licenced under Creative Commons Attribution-Share Alike 4.0.

the enzymes involved in transcription and translation. Other aspects of archaeal biochemistry are unique, such as their reliance on ether lipids in their cell membranes, including archaeols. Archaea are the target of a number of viruses in a diverse virosphere distinct from bacterial and eukaryotic viruses.

Archaea and bacteria are generally similar in size and shape. The first observed archaea were extremophiles, living in extreme environments, such as hot springs and salt lakes with no other organisms. Improved molecular detection tools led to the discovery of archaea in almost every habitat, including soil, oceans, and marshlands. Archaea are particularly numerous in the oceans, and the archaea in plankton may be one of the most abundant groups of organisms on the planet.

Salt-tolerant archaea (the Haloarchaea) use sunlight as an energy source, and other species of archaea fix carbon, but unlike plants and cyanobacteria, no known species of archaea does both.

The electron density of archaea that grow under conditions of high temperature and low pH, such as Sulfolobus, may be rather low (approximately 30 mV).

Viruses and Phages

These are not cellular organisms and thus outside my remit to describe electricity in living cells. Nevertheless they have an impact on cellular organisms causing, for instance, pandemics of influenza and

Covid-19. Therefore I thought I ought to say something about them. Viruses are submicroscopic infectious agents that replicate only inside the living cells of an organism. Viruses infect all types of life forms, from animals and plants to microorganisms, including bacteria and archaea. Many viruses are surrounded by a continuous bilayer membrane studded with viral proteins. Its purpose is to protect the genome-containing virus nucleocapsid from damage, and to facilitate entry of the nucleocapsid into a host cell. Viral membrane proteins attach the virus to the host cell and promote fusion between viral and host cell membranes. Enveloped viruses form by a budding process, involving the wrapping up of newly synthesised nucleocapsids with regions of host cell membranes containing host cell lipids but exclusively viral proteins. Host cell machinery and processes are co-opted by enveloped viruses for use at every stage of these viral events (Lenard, 2008).

Phages are viruses that infect and replicate within bacteria and archaea. Phages also attack living cells, and as with viruses, might be beneficial in possible treatment of bacterial disease, especially when the pathogen of concern is antibiotic resistant.

Electric currents have some effects on any live organism, including viruses, e.g., HIV-1 (Kumagai, 2011). This finding was motivated by a search for treatments for AIDS. Upon binding to receptor-bearing target cells, viruses cause cell membrane potential changes. Epstein-Barr Virus causes a biphasic membrane potential change in receptor-bearing B lymphocytes but not receptor-negative T lymphocytes, as measured by flow cytometry or cyanine dye uptake. A safer approach to this type of experiment is to select a non-vital virus and conduct it with low electric currents and evaluate how it was affected. Regarding the safety of the research team and availability, the non-vital herpes simplex virus seems suitable. The efforts now being made to detect and quantify viruses electrically (Al Ahmad, 2014) suggests that viruses have responding electrical properties.

Light-emitting organisms

Fireflies are soft-bodied beetles (arthropods) that are commonly also called glow worms or lightning bugs for their conspicuous use of bioluminescence during twilight to attract mates or prey. Fireflies produce a "cold light", with no infrared or ultraviolet frequencies. This chemically produced light from the lower abdomen may be yellow, green, or pale red, with wavelengths from 510–670 nanometres. Some species such as the dimly glowing "blue ghost" of the Eastern U.S. are commonly thought to emit blue light, although this is a false perception of their truly green emission light, due to the Purkinje effect, i.e., the tendency for the peak luminescence sensitivity of the human eye to shift toward the blue end of the colour spectrum at low illumination levels as part of dark adaptation. In consequence, reds will appear darker relative to other colours as light levels decrease. Fireflies produce a chemical reaction inside their bodies that allows them to light up. This type of light production is called bioluminescence.

Bioluminescence requires the presence of certain organic chemicals. As we know from Scudder (2013), there are electron flows in organic chemicals, which make me wonder whether this is a transformation of electrical energy (electron flow) into light energy (photons), with consumption of the former, as is the case with man-made lighting equipment. However, there are other factors that vary with the organism concerned. Bioluminescence is also found in many marine organisms such as bacteria, algae, jellyfish, worms, crustaceans, sea stars, fish, and sharks, to name just a few. In fish alone, there are about 1,500 known species that show luminescence. In some cases, animals take in bacteria or other bioluminescent creatures to gain the ability to light up. Generally, light emission occurs in many different groups of organisms, but it involves only two major processes with differing mechanisms of light production, namely bioluminescence and fluorescence. Some organisms have developed the ability to utilise both bioluminescence and fluorescence as tandem systems.

When oxygen combines with calcium, adenosine triphosphate (ATP) and the chemical luciferin in the presence of luciferase, a bioluminescent enzyme, light is produced. In fireflies' light-producing cells (photocytes) of the light organ, mitochondria are clustered in the cell periphery (Aprille, 2004), positioned between the tracheolar air supply and the oxygen-requiring bioluminescent reactants which are sequestered in more centrally localised peroxisomes (a membrane-bound organelle, formerly known as a microbody, found in the

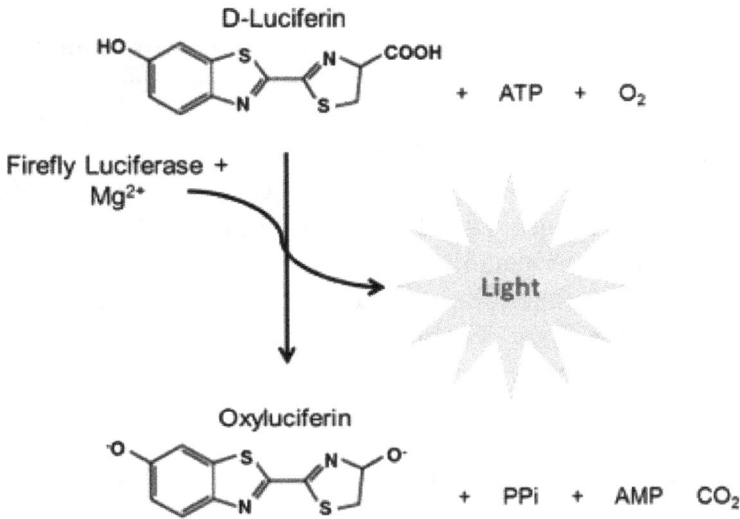

Figure 95: The ATP assay using the D-Luciferin/Firefly luciniferase[+] reaction. Source: The luciferase ATP assay kit instructions.

cytoplasm of virtually all eukaryotic cells). This clustering of mito-chondria suggests a high electron generation and ATP production. When I was in gainful employment and had access to laboratory facili-ties, a colleague used to perform an ATP assay (Figure 95) for me using a similar process *in vitro*. Light is emitted because the reaction forms oxyluciferin in an electronically excited state.

Probably, in the intact organism, Ca^{2+} rather than Mg^{2+} is the co-factor. Excitation-luminescence coupling has been studied by Bil-baut (*1980*) in the polynoid worm *Acholoe astericola*. In the photo-genic epithelial cells (photocytes), the increase of the stimulus strength elicited an action potential specifically correlated with a light flash. This membrane response begins by a fast overshooting Ca-dependent spike potential followed by an Na-dependent secondary depolari-sation (electron outflow). The excitation-luminescence coupling is dependent on Ca^{2+} entry into the photocytes. Thus, we have something (Figure 95) that looks very similar to mechanisms of excitation-contraction coupling in muscle, in which depolarisation (electron outflow) opens a Ca^{2+} channel which triggers the ATP/heavy meromyosin reaction.

Aequorin is a bioluminescent protein obtained from the jellyfish (*Aequorea victoria*). It is a 22 kilo-Dalton protein complex that contains bound oxygen, the luciferin coelenterazine, and three calcium-binding sites. The *Aequorea victoria* green fluorescent protein is a useful visual marker enabling resolution of single action potentials and can reveal subthreshold activity. Recombinant aequorin and assays based on aequorin labels have been developed in the past.

When it comes to aequorin rather than luciferase, it is evident that coelenterazine is membrane permeable, and can be used to facilitate the reassembling of the aequorin complex *in vivo*. Coelenterazine is oxidised and illuminates blue light at 466 nm when Ca^{2+} binds to the complex. The luminescence intensity is correlated to the Ca^{2+} concentration, and was used to assay Ca^{2+} in studies of excitation-contraction coupling in muscle (Allen & Kurihara, *1980*). A new generation of intracellular calcium indicators have since been produced, in which engineered variants of green fluorescent protein are used to probe their ionic environment using intramolecular fluorescence resonance energy transfer.

Final Conclusion: Living cells show electricity. Electricity = electrons moving.

Epilogue

In the preface, I indicated how theories about the nature of the universe were sequentially replaced, so that now we are in the era of space-time and quantum fields. The latter appeared to me to be applicable to theories about physiological processes (including electrophysiology that I was asked to write about), but I found very few such interpretations in my literature searches.

Reminder: **Theories cannot be proved. Science progresses through disproof of theories and the adoption of newer and hopefully better theories.**

1. The first established theory considered, that membrane potential was determined by the extracellular/intracellular electrolyte distribution, was easy to disprove after I realised that the red blood cell (conventional polarity -9 mV) had the same electrolyte distribution as cardiac ventricular muscle (at -85 mV).

I do not like the arbitrary human assignation of negativity to electricity ("negative" electrons moving), so I adopted Scudder's concept of electron density (Scudder, 2013).

I do not like the term "membrane potential" which implies that the electricity of a cell is determined by the cell membrane activity. That is impossible because lipid bilayers cannot generate electricity. "Trans-membrane potential difference" is more acceptable.

Two alternative theories were required for future physiologists to try and accept the disproval of theory number 1:

(A) The first one I came across was, **"The distribution of electrolytes by living cells is genetically determined."**

(B) The second one was, **"The conventional polarity making the intracellular voltage negative is determined by electron generation within the cell. As the organelle with the most negative voltage was the mitochondrion, I postulate that electron generation takes place in mitochondria during oxidative phosphorylation."**

There are no mitochondria in bacteria, but oxidative phosphorylation does take place, generating enough electrons to create the most negative trans-membrane potential differences that I have found in my literature searches of electrophysiological difference between different species (comparative electrophysiology).

As the electron is a sub-atomic particle, i.e., a quantum, the new theories (A&B) take electrophysiology into the global theory of quantum fields.

2. The second established theory considered was that depolarisation and repolarisation are achieved by cation (positively charged ion) currents. This I considered unsatisfactory because: [a] One of these currents is called "funny current". If investigators are unable to find the cation that carries the cardiac sino-atrial pacemaker current, the logical conclusion is that it does not exist. [b] In the steady state, ion current from the extracellular to the intracellular compartment must be balanced by an equal quantity transfer of that ion back to the exterior; ion current from the intracellular to the extracellular compartment must be balanced by an equal quantity transfer of that ion to the interior, i.e., this theory (with the exception of the calcium ion) does not satisfy the ionic balance necessity. [c] Whereas depolarisation in animal cells is attributed to cation inflow, depolarisation in land plants is attributed to anion (negatively charged ion) outflow. [d] Jellyfish have no sodium channel. [e] Nobody to my knowledge thus far *measured* the travel of these ions from and to the sites postulated by the current theories.

I mooted the need for calcium ion balance as long ago as 1979 (Noble, *1979*); the mechanism by which this is achieved has now been documented by Eisner (*2012*).

While these considerations may not constitute absolute disproof, there was enough discouragement for me to consider the need for a better theory for future physiologists to try and disprove.

The theory is: **Depolarisation is an outflow of electrons, from the cytoplasmic interior of cell outwards through the cell membrane to the extracellular compartment. Repolarisation is a flow of electrons generated within the cell substance (bacteria) or by the mitochondria in aerobic cells during the process of oxidative phophorylation.**

With this theory, the problem of ionic imbalance does not exist.

It is a one-way system: electron generation to electron flow to depolarisation to triggering function to consumption and dissipation to heat.

I like this theory because it seems to be almost universal amongst aerobic organisms, so the reader may be critical of my neglect of anaerobic organisms; that would need a separate work by someone specialising in that huge field of endeavour.

The reader may be disappointed if I have not discussed their favourite organism of interest and study, but this work was not intended to be an encyclopaedia, so I picked and chose examples of the possible application of various theories.

Another common occurrence in aerobic organisms is that depolarisation allows entry into the cell of a tiny amount of Ca^{2+} that is necessary for cells to achieve their various functions. This has to be extremely carefully controlled as we do not want to end up in the white cliffs of Dover as fossils of organisms killed by calcium overload. As I did not want to get into this subject in a work about electrophysiology, I have simply followed the conventional custom calling this, "opening of the Ca^{2+} channel".

3. The third established theory considered in this work is that electrons are not involved in muscle contraction. Contraction is conventionally considered to occur solely by mechanical interactions stimulated by Ca^{2+} and ATP. Cross bridges do not cross anything but seem to behave as nano-muscles that grab the thin filaments and yank them. If true, the thin filaments would all be pulled into the centre of the sarcomere.

As long ago as 1989, I thought that this theory could not explain the sliding of muscle thin filaments between the myosin thick filaments (Iwazumi & Noble, 1989).

Alternative theories exist which future physiologists ought to try and disprove, of which the following is an example.

A quantum field of heavy meromyosin, when activated, draws the thin filaments straight in toward the centre of the sarcomere.

This theory is a beautiful example of the modern concept that the universe consists of space-time and quantum fields (see preface).

Electrons seem to be involved in all living matter, even in excrement from which they can extract energy (pee power & poo power) as well as many useful molecules.

References

(*listed by alphabetical order of first author.*)

Abbott BC, Aubert XM (1952) The force exerted by active striated muscle during and after change of length. *J Physiol* 117:77–86.

Al Ahmad M, Mustafa F, Ali LM, Rizvi TA (2014) Virus detection and quantification using electrical parameters. *Sci Rep* 4:6831.

Alberts B, Johnson A, Lewis J, Raff M, Roberts K, Walter P (2002) Ion channels and the electrical properties of membranes. In: *Molecular Biology of the Cell* (4th edition), New York: Garland Science.

Allen DG, Kurihara S (1980) Calcium transients in mammalian ventricular muscle. *Eur Heart J* 1:5–15.

Anastatasaki E, Drizou F, Milonas PG (2018) Electrophysiological and oviposition responses of Tuta absoluta females to herbivore-induced volatiles in tomato plants. *J Chem Ecol* 44:288–298.

Andersen JL, MacMillan HA, Overgaard J (2015) Muscle membrane potential and insect chill coma. *J Exp Biol* 218:2492–2495.

Andersen O (2013) Cellular electrolyte metabolism. In: *Encyclopedia of Metalloproteins*, New York: Springer, pp. 580–587.

Aprille JR, Lagace CJ, Modica-Napolitano J, Trimmer BA (2004) Role of nitric oxide and mitochondria in control of Firefly flash. *Integr Comp Biol* 44:213–219.

Arlock P, Noble MIM, Wohlfart B (1981) Cardiac cell membrane repolarisation is required for the onset of mechanical restitution in papillary muscle. *Acta Physiol Scand* 142:113–118.

Arlock P, Wohlfart B, Noble MIM (1992) Effects of transient changes in membrane potential on twitch force in ferret papillary muscle. *Ann NY Acad Sci* 639:456–459.

Bailey DM (2019) Making sense of oxygen; quantum leaps with 'physics-iology'. *Exp Physiol* 104:453–457.

Bartoli A, Cross BA, Guz A *et al.* (1974) The effect of carbon dioxide in the airways and alveoli on ventilation: A vagal reflex studied in the dog. *J Physiol* 240:91–109.

Brault M, Amiar Z, Pennarun A-M *et al.* (2004) Plasma membrane depolarisation induced by abscisic acid in Arabidopsis suspension cells involves reduction of proton pumping in addition to anion channel activation, which are both Ca^{2+} dependent. *Plant Physiol* 135:231–243.

Behar J (2013) Physiology and pathophysiology of the biliary tract: The gallbladder and sphincter of Oddi — a review. *Int Sch Res Notices* 2013:837630.

Benarroch JM, Asally M (2020) The microbiologist's guide to membrane potential dynamics. *Trends Microbiol* 28:304–314.

Berecki G, Wilders R, de Jonge B, van Ginneken ACG, Verkerk AO (2010) Re-evaluation of the action potential upstroke velocity as a measure of the Na^+ current in cardiac myocytes at physiological conditions. *PLoS One* 5:e05772.

Bilbaut A (1980) Excitable epithelial cells in the bioluminescent scales of a polynoid worm: Effects of various ions on the action potentials and on the excitation-luminescence coupling. *J Exp Biol* 88:219–238.

Bird J (2014) *Electrical and Electronic Principles and Technology* (5th edition), London: Routledge.

Bot CT, Prodan B (2010) Quantifying the membrane potential during *E. coli* growth stages. *Biophys Chem* 146:133–137.

Boyett MR, Hart G, Levi AJ, Roberts A (1987) Effects of repetitive activity on developed force and intracellular sodium in isolated sheep and dog Purkinje fibres. *J Physiol* 388:295–322.

Brading AF (2006) Spontaneous activity of lower urinary tract smooth muscles: Correlation between ion channels and tissue function. *J Physiol* 570:13–22.

Bradley GW, Noble MIM, Trenchard D (1976) The direct effect on pulmonary stretch receptor discharge produced by changing lung carbon dioxide concentrations in dogs on cardiopulmonary bypass and its action on breathing. *J Physiol* 261:359–373.

Bulbring E (1985) Correlation between membrane potential, spike discharge and tension in smooth muscle. *J Physiol* 128:200–221.

Calderon JC, Bolanos P, Caputo C (2014) The excitation–contraction coupling mechanism in skeletal muscle. *Biophys Rev* 6:133–160.

Cech DJ, Martin S (2012) Sensory system changes. In: *Functional Movement Development Across the Life Span* (3rd edition), Philapelphia: Saunders.

Chen Y, Guzic S, Sumner JP *et al.* (2011) Magnetic manipulation of actin orientation, polymerisation, and gliding on myosin using superparamagnetic iron oxide particles. *Nanotechnology* 22:065101.

Cole M, Lindeque P, Fileman E, Halsband C *et al.* (2013) Microplastic ingestion by zooplankton. *Environ Sci Technol* 47:6646–6655.

Costa RR, Varanda WA, Franci CR (2010) A calcium-induced calcium release mechanism supports luteinising hormone-induced testosterone secretion in mouse Leydig cells. *Am J Physiol Cell Physiol* 299:C316–C323.

Curtin NA, Woledge RC (1978) Energy changes and muscular contraction. *Physiol Rev* 58:690–761.

Daniels M, Noble MIM, ter Keurs HEDJ, Wohlfart B (1984) Velocity of sarcomere shortening in rat cardiac muscle: Relationship of force, sarcomere length, calcium and time. *J Physiol* 355:367–381.

Dean PM, Matthews EK (1970) Glucose-induced electrical activity in pancreatic islet cells. *J Physiol* 210:255–264.

Dean PM, Matthews EK (1970) Electrical activity in pancreatic islet cells: Effect of ions. *J Physiol* 210:265–275.

DeCoursey TE (2000) Hypothesis: Do voltage-gated H^+ channels in alveolar epithelial cells contribute to CO_2 elimination by the lung? *Am J Physiol Cell Physiol* 278:C1–C10.

De la Fuente IM, Bringas C, Malaina I *et al.* (2019) Evidence of conditioned behavior in amoebae. *Nat Commun* 10:3690.

Draper MH, Weidmann S (1951) Cardiac resting and action potentials recorded with an intracellular electrode. *J Physiol* 115:74–94.

Edman KAP, Johannsson M (1976) The contractile state of rabbit papillary muscle in relation to stimulation frequency. *J Physiol* 254:565–581.

Edman KAP, Elzinga G, Noble MIM (1978) Enhancement of mechanical performance by stretch during tetanic contractions of vertebrate skeletal muscle fibres. *J Physiol* 281:139–155.

Eisner D, Bode E, Venetucci L, Trafford A (2013) Calcium flux balance in the heart. *J Mol Cell Cardiol* 58:110–117.

Elzinga G, Lab MJ, Noble MIM *et al.* (1981) The action potential and contractile response of the intact heart related to the preceding interval and the preceding beat in the dog and cat. *J Physiol* 314:481–500.

Fabiato A, Fabiato F (1975) Contractions induced by a calcium-triggered release of calcium from the sarcoplasmic reticulum of single skinned cardiac cells. *J Physiol* 249:469–495.

Felle H, Porter JS, Slayman CL, Kaback HR (1980) Quantitative measurements of membrane potential in *Escherichia coli*. *Biochemistry* 19:3585–3590.

Feynman R (2011) *The Feynman Lectures on Physics*, New York: Basic Books.

Fingerle J, Grasmann D (1982) Electrical properties of the plasma membrane of microplasmodia of *Physarum polycephalum*. *J Membr Biol* 68:67–77.

Ford LE, Huxley AF, Simmons RM (1977) Tension responses to sudden length change in stimulated frog muscle fibres near slack length. *J Physiol* 269:441–515.

Fowler CJ, Griffiths D, de Groat WC (2008) The neural control of micturition. *Nat Rev Neurosci* 9:453–466.

Friedrich T, Steinmuller K, Weiss H (1995) The proton-pumping respiratory complex I of bacteria and mitochondria and its homologue in chloroplasts. *FEBS Lett* 26:107–111.

Gillian RE, Senthil Kumar VS, O'Neall-Hennessey E *et al.* (2013) X-ray solution scattering of squid heavy meromyosin: Strengthening the evidence for an ancient compact off state. *PLoS One* 17:e81994.

Gradmann D, Boyd CM (1995) Membrane voltage of marine phytoplankton, measured in the diatom *Coscinodiscus radiatus*. *Mar Biol* 123:645–650.

Gordon DG (2010) *The Secret World of Slugs and Snails: Life in the Very Slow Lane*, Seattle: Sasquatch Books.

Guz A, Noble MIM (1970) The role of vagal inflation reflexes in man and other animals. In: *Breathing: Hering-Breuer Centenary Symposium*, London: J. & A. Churchill.

Guz A, Noble MIM, Eisele JH, Trenchard D (1971) The effect of lung deflation on breathing in man. *Clin Sci* 40:451–461.

Hagawara N, Irisawa H, Kameyama M (1988) Contribution of two types of calcium currents to the pacemaker potentials of rabbit sino-atrial node cells. *J Physiol* 395:233–253.

Hallen J (2012) Troponin for the estimation of infarct size: What have we learned? *Cardiology* 121:204–212.

Halliwill JR, Morgan BJ, Charkoudian N (2003) Peripheral chemoreflex and baroreflex interactions in cardiovascular regulation in humans. *J Physiol* 552:295–302.

Harrison SM, McCall E, Boyett MR (1982) The relationship between contraction and intracellular sodium in rat and guinea pig ventricular myocytes. *J Physiol* 449:517–550.

Hashitani H, Brading AF (2003) Electrical properties of detrusor smooth muscles from the pig and human urinary bladder. *Br J Pharmacol* 140:146–158.

Hayashi F, Sinclair JD (1991) Respiratory patterns in anaesthetised rats before and after anemic decerebration. *Respir Physiol* 84:61–76.

Higinbotham N (1973) Electropotentials of plant cells. *Ann Rev Plant Physiol* 24:25–46.

Hill R (1937) Oxygen evolved by isolated chloroplasts. *Nature* 139:881–882.

Hill R (1939) Oxygen produced by isolated chloroplasts. *Proc Royal Soc B* 127:192–210.

Hill R, Scarisbrick R (1940) Production of oxygen by illuminated chloroplasts. *Nature* 146:61–62.

Hilton SM (1959) A peripheral arterial conducting mechanism underlying dilatation of the femoral artery and concerned in functional vasodilatation in skeletal muscle. *J Physiol* 149:93–111.

Hodgkin AL, Huxley AF (1952) A quantitative description of membrane current and its application to conduction and excitation in nerve. *J Physiol* 117:500–544.

Holash RJ, MacIntosh BR (2019) A stochastic simulation of skeletal muscle calcium transients in a structurally realistic sarcomere model using MCell. *PloS Comput Biol* 15:e1006712.

Hume JR, Grant AO (2012) Agents used in cardiac arrhythmias. In: *Basic & Clinical Pharmacology* (13th edition), New York: McGraw-Hill Education.

Huxley HE (1966) The fine structure of striated muscle and its functional significance. *Harvey Lect* 60:85–118.

Huxley HE, Hansen J (1954) Changes in the cross-striations of muscle during contraction and stretch of muscle and their structural interpretation. *Nature* 173:973–976.

Huxley AF, Niedergerke R (1954) Structural changes in muscle during contraction: Interference microscopy of living muscle fibres. *Nature* 173:971–973.

Ikeda K, Kawakami K, Onimaru H *et al.* (2017) The respiratory control mechanisms in the brainstem and spinal cord: Integrative views of the neuroanatomy and neurophysiology. *J Physiol Sci* 67:45–62.

Iwazumi T (1970) A new field theory of muscle contraction. PhD Thesis, University of Pennsylvania. Microfilms Inc, Ann Arbor, Michigan.

Iwazumi T (1989) Molecular mechanism of muscle contraction. *Physiol Chem Phys Med NMR* 21:187–219.

Iwazumi T, Noble MIM (1989) An electrostatic mechanism of muscular contraction. *Int J Cardiol* 24:267–275.

Iwazumi T (1988) The mechanism of the length-tension relation in cardiac muscle of *Rana Catesbeiana*. In: *Starling's Law of the Heart Revisited*, Dordrecht: Kluwer Academic Publishers, pp. 18–27.

Jalife J, Hamilton AJ, Lamanna VR, Moe GK (1980) Effects of current flow on pacemaker activity of the isolated kitten sinoatrial node. *Am J Physiol* 238:H207–H316.

Jimenez JL, Crone SPG, Fogh E *et al.* (2021) A quantum magnetic analogue to the critical point of water. *Nature* 592:370–375.

Johansson B, Mellander S (1975) Static and dynamic components in the vascular myogenic response to passive changes in length as revealed by electrical and mechanical recordings from the rat portal vein. *Circ Res* 36:76–83.

Josefsson J-O (1966) Some bioelectrical properties of *Amoeba proteus*. *Acta Physiol Scand* 66:395–405.

Katz AI (1982) Renal Na-K-ATPase: Its role in tubular sodium and potassium transport. *Am J Physiol* 242:F207–F219.

Kawa K (1987) Existence of calcium channels and intercellular couplings in the testosterone-secreting cells of the mouse. *J Physiol* 393:647–666.

Keller T, Zeller T, Peetz D *et al.* (2009) Sensitive troponin I assay in early diagnosis of acute myocardial infarction. *N Engl J Med* 361:868–877.

Kelly RF, Snow HM (2011) The effect of arterial wall stress on the incremental elasticity of a conduit artery. *Acta Physiol (Oxf)* 202:1–9.

Kelly R, Ruane O'Hora T, Noble MIM *et al.* (2006) Differential inhibition by hyperglycaemia of shear stress- but not acetylcholine-mediated dilatation in the iliac artery of the anaesthetised pig. *J Physiol* 573:133–145.

Kentish JC, ter Keurs HEDJ, Ricciardi L, Bucx JJJ, Noble MIM (1986) Comparison between the sarcomere length-force relations of intact and skinned trabeculae from rat right ventricle. Influence of calcium concentrations on these relations. *Circ Res* 58:755–768.

Kobayasi S, Totsuka T (1975) Electric birefringence of myosin subfragments. *Biochim Biophys Acta Bioenerg* 376:375–385.

Kronhaus KD, Spear JF, Moore EN, Kline RP (1978) Sinus node extracellular potassium transients following vagal stimulation. *Nature* 275:322–324.

Kumagai E, Tominaga M, Harada S (2011) Sensitivity to electrical stimulation of human immunodeficiency virus type 1 and MAGIC-5 cells. *AMB Express* 1:23.

Kupisz K, Dziubinska H, Trecacz K (2017) Generation of action potential-type changes in response to darkening and illumination as indication of the plasma membrane proton pump status in *Marchantia polymorpha*. *Acta Physiol Plant* 39:82.

Lambers H *et al.* (2021) Photosynthesis Biology. University of Western Australia, Crawley, Western Australia.

Lan L, Oswald SJ, Gnu H *et al.* (2008) Mechanical properties of actin stress fibres in living cells. *Biophys J* 95:6060–6071.

Langer GA (1985) The effect of pH on cellular and membrane calcium binding and contraction of myocardium. A possible role for sarcolemmal phospholipid in EC coupling. *Circ Res* 57:374–382.

Lenard J (2008) Viral membranes. In: *Encyclopedia of Virology* (3rd edition), Cambridge: Academic Press, pp. 308–314.

Lin Y-CJ, Spencer AN (2001) Calcium currents from jellyfish striated muscle cells: Preservation of phenotype, characterisation of currents and channel localisation. *J Exp Biol* 204:3717–3726.

Linari M, Reedy MK, Reedy MC *et al.* (2004) Ca-activation and stretch-activation in insect flight muscle. *Biophys J* 87:1101–1111.

Mabuchi K (1991) Heavy-meromyosin-decorated actin filaments: A simple method to preserve actin filaments for rotary shadowing. *J Struct Biol* 107:22–28.

Mansfield PB, McDonald RH (1965) Some metabolic aspects of paired pacing of the heart. *Bull NY Acad Med* 41:700–711.

Markos F, Ruane O'Hora T, Noble MIM (2013) What is the mechanism of flow mediated arterial dilatation? *Clin Exp Pharmacol Physiol* 40, 489–494.

Markos F, Ruane O'Hora T, Wainwright CL *et al.* (2012) Dependence of smooth muscle tone upon pulsatility in the iliac artery of the anaesthetised pig. *Pflugers Arch* 463:679–684.

Martini M, Canella R, Rubbini G, Riccardo F, Rossi ML (2015) Sensory transduction at the frog semicircular canal: How hair cell membrane potential controls junctional transmission. *Front Cell Neurosci* 9:235.

Matthews EK, Petersen OH (1973) Pancreatic acinar cells: Ionic dependence of the membrane potential and acetylcholine-induced depolarisation. *J Physiol* 231:283–295.

Mellander S, Arvidsson S (1974) Possible 'dynamic' component in the myogenic vascular response related to pulse pressure distension. *Acta Physiol Scand* 90:283–285.

Meyer DJ, Jones CW (1973) Oxidative phosphorylation in bacteria which contain different cytochrome oxidases. *Eur J Biochem* 36:144–151.

Mezzano V, Pellman J, Sheikh F (2014) Cell junctions in the specialised conduction system of the heart. *Cell Commun Adhes* 21:149–159.

Milo R, Philips R (2015) What are the concentrations of different ions in cells? In: *Cell Biology by the Numbers*, New York: Garland Science. Retrieved 2020 from http://book.bionumbers.org/what-are-the-concentrations-of-different-ions-in-cells/.

Mithofer A, Mazars C (2002) Aequorin-based measurements of intracellular Ca^{2+} signatures in plant cells. *Biol Proced Online* 4:105–118.

Miyashita T, Gess RW, Tietjen K, Coates MI (2021) Non-ammocoete larvae of Palaeozoic stem lampreys. *Nature* 591:408–412.

Mochizuki S, Vink H, Hiramatsu O *et al.* (2003) Role of hyaluronic acid glycosaminoglycans in shear-induced endothelium-derived nitric oxide release. *Am J Physiol Heart Circ Physiol* 285:H722–H726.

Moore JC (1984) The Golgi tendon organ: A review and update. *Am J Occup Ther* 38:227–236.

Morel JE, Garrigos M (1982) Dimerisation of the myosin heads in solution. *Biochemistry* 21:2679–2686.

Lennarz W, Lane M (2013) *Encyclopedia of Biological Chemistry* (2nd edition), New York: Academic Press.

Nakajima S, Onodera K (1969) Membrane properties of the stretch receptor neurones of crayfish with particular reference to mechanisms of sensory adaptation. *J Physiol* 200:161–185.

Natochin YV (2010) From quantum to integrative physiology. *Ross Fiziol Zh Im I M Sechenova* 96:1043–1061.

Nattie E, Li A (2012) Central chemoreceptors: Locations and functions. *Compr Physiol* 2:221–254.

Nichols CG (2006) KATP channels as molecular sensors of cellular metabolism. *Nature* 440:470–476.

Noble MIM (1979) The calcium cycle. In: *The Cardiac Cycle*, Oxford: Blackwell Scientific Publications.

Noble MIM, Stubbs J, Trenchard D, Else W, Eisele JH, Guz A (1972) Left ventricular performance in the conscious dog with chronically denervated heart. *Cardiovasc Res* 6:457–475.

Noble MIM, Arlock P, Wohlfart B, Drake-Holland AJ (2006) The beat-to-beat decay of cardiac contractility from highly potentiated levels is bi-exponential. *J Biomech* 39:2657–2664.

Noble MIM, Else W (1972) Re-examination of the applicability of the Hill model of muscle to cat myocardium. *Circ Res* 31:580–589.

Noble MIM, Arlock P, Gath J *et al.* (1993) Mechanisms of excitation-contraction coupling using the principle of transient perturbation. *Cardiovasc Res* 27:1758–1765.

Offer G, Baker H, Baker L (1972) Interaction of monomeric and polymeric actin with myosin subfragment 1. *J Mol Biol* 66:435–444.

Paredes RM, Etzler JC, Watts LT *et al.* (2008) Chemical calcium indicators. *Methods* 46:143–151.

Philips RJ, Powley TL (2000) Tension and stretch receptors in gastrointestinal smooth muscle: Re-evaluating vagal mechanoreceptor electrophysiology. *Brain Res Rev* 34:1–26.

Philips PJ, Gwathmey JK, Feldman MD *et al.* (1990) Post-extrasystolic potentiation and the force-frequency relationship: Differential augmentation of myocardial contractility in working myocardium from patients with end-stage heart failure. *J Mol Cell Cardiol* 22:99–110.

Picht E, Zima AV, Shannon TR *et al.* (2011) Dynamic calcium movement inside cardiac sarcoplasmic reticulum during release. *Circ Res* 108:847–856.

Rathi A (2014) "Jellyfish are the most energy-efficient swimmers, new metric confirms". *Ars Technica*. Archived from the original on 3 November 2014. Retrieved 3 December 2014.

Redman RS (1994) Myoepithelium of salivary glands. *Microsc Res Tech* 27:25–45.

Reedy MK (1968) Ultrastructure of insect flight muscle. I. Screw sense and structural grouping in the rigor cross-bridge lattice. *J Mol Biol* 31:155–176.

Rovelli C (2014) *Reality is Not What It Seems: The Journey to Quantum Gravity*, New York: Riverhead Books.

Roper S (1983) Regenerative impulses in taste cells. *Science* 220:1311–1312.

Ruegg JC (1967) ATP-driven oscillation of glycerol-extracted insect fibrillar muscle: Mechano-chemical coupling. *Am Zoologist* 7:457–464.

Saenger W (1987) Structure and dynamics of water surrounding biomolecules. *Ann Rev Biophys Biophys Chem* 16:93–114.

Sanchez C, Berthier C, Allard B *et al.* (2018) Tracking the sarcoplasmic reticulum membrane voltage in muscle with a FRET biosensor. *J Gen Physiol* 150:1163–1177.

Sawanobori T, Takanashi I, Hiraoka M, Iida Y, Kamisaka K, Maezawa H (1989) Electrophysiological properties of isolated rat liver cells. *J Cell Physiol* 139:580–585.

Scudder PH (2013) *Electron Flow in Organic Chemistry: A Decision-Based Guide to Organic Mechanisms* (2nd edition), New Jersey: John Wiley & Sons, Inc.

Sedaghat G, Ghafoori E, Waks JW *et al.* (2016) Quantitative assessment of vectorcardiographic loop morphology. *J Electrocardiol* 49:154–163.

Spector T (2020) *Spoon-Fed: Why Almost Everything We've Been Told about Food is Wrong*, London: Vintage Publishing.

Stojilkovic SS (2012) Molecular mechanisms of pituitary endocrine cell calcium handling. *Cell Calcium* 51:212–221.

Stubbs P, Collinson P, Moseley D, Greenwood T, Noble MIM (1996a) Prospective study of the role of cardiac troponin T in patients admitted with unstable angina. *BMJ* 313:262–264.

Stubbs P, Collinson P, Moesely D, Greenwood T, Noble MIM (1996b) Prognostic significance of admission troponin T concentrations in patients with myocardial infarction. *Circulation* 94:1291–1297.

Suchodolski J, Krasowska A (2019) Plasma membrane potential of *Candida albicans* measured by Di-4-ANEPPS fluorescence depends on growth phase and regulatory factors. *Microorganisms* 7:110.

Taylor AR (2009) A fast Na^+/Ca^{2+}-based action potential in a marine diatom. *PloS One* 4:e4966.

Tanaka Y, Noguchi Y, Yalikun Y *et al.* (2017) Earthworm muscle driven bio-micropump. *Sens Actuators B Chem* 242:1186–1192.

Templeton GH, Mitchell JH, Ecker RR *et al.* (1970) A method for measurement of dynamic compliance of the left ventricle in dogs. *J Appl Physiol* 29:742–745.

Templeton GH, Ecker RR, Mitchel JH (1972) Left ventricular stiffness during diastole and systole: The influence of changes in volume and inotropic state. *Cardiovasc Res* 6:95–100.

Templeton GH, Nardizzi LR (1974) Elastic and viscous stiffness of the canine left ventricle. *J Appl Physiol* 36:123–127.

Ter Keurs HEDJ, Noble MIM (1988) *Starling's Law of the Heart Revisited*, Dordrecht: Kluwer Academic Publishers.

Ter Keurs HEDJ, Gao WD, Bosker H *et al.* (1990) Characterisation of decay of frequency induced potentiation and post-extrasystolic potentiation. *Cardiovasc Res* 24:903–910.

Torsoli A (1971) The function of biliary "sphincters". *JR Coll Surg Edinb* 16:270–273.

Ullrich CI, Novacky AJ (1991) Electrical membrane properties of leaves, roots, and single root cap cells of susceptible *Avena sativa*. *Plant Physiol* 95:675–681.

Vallverdu J, Castro O, Mayne R *et al.* (2018) Slime mould: The fundamental mechanisms of biological cognition. *Biosystems* 165:57–70.

Vestergaard M, Nøhr-Meldgaard K, Ingmer H (2018) Multiple pathways towards reduced membrane potential and concomitant reduction in aminoglycoside susceptibility in *Staphylococcus aureus. Int J Antimicrob Agents* 51:132–135.

Vinberg F, Wang T, De Maria A, Zhao H, Bassnett S, Chen J, Kefalov VJ (2017) The $Na^+/Ca^{2+},K^+$ exchanger NCKX4 is required for efficient cone-mediated vision. *eLife* 6:e24550.

Vodeneev V, Akinchits E, Sukhov V (2015) Variation potential in higher plants: Mechanisms of generation and propagation. *Plant Signal Behav* 10:e1057365.

Volkov EM, Nurullin LF, Frosin VN (2001) Effect of cholinergic agonists on resting membrane potential of earthworm body wall muscle cells. *Bull Exp Biol Med* 131:397–398.

Volkov EM, Nurullin LF, Svandova I *et al.* (2000) Participation of electrogenic Na^+-K^+-ATPase in the membrane potential of earthworm body wall muscles. *Physiol Res* 49:481–484.

Wakin KG (1971) Passive role of the bile duct system in the delivery of bile into the intestine. *Surg Gynecol Obstet* 133:826–829.

Wang D, Grillner S, Wallen P (2013) Calcium dynamics during NMDA-induced membrane potential oscillations in lamprey spinal neurones — contribution of L-type calcium channels (CaV1.3). *J Physiol* 591:2509–2521.

Weidmann S (1951) Effect of current flow on the membrane potential of cardiac muscle. *J Physiol* 115:227–236.

Weinbaum S, Zhang X, Han Y, Vink H, Cowin SC (2003) Mechanotransduction and flow across the endothelial glycocalyx. *Proc Natl Acad Sci USA* 100:7988–7995.

White CR, Frangos JA (2007) The shear stress of it all: The cell membrane and mechanochemical transduction. *Philos Trans R Soc London B Biol Sci* 362:1459–1467.

Wohlfart B (1979) Relationships between peak force, action potential duration and stimulus interval in rabbit myocardium. *Acata Physiol Scand* 106:395–409.

Wolynes PG (2009) Some quantum weirdness in physiology. *Proc Nat Acad Sci USA* 106:17247–17248.

Woods CM, Mawe GM, Toouli J *et al.* (2005) The sphincter of Oddi: Understanding its control and function. *Neurogastroenterol Motil 17 Supp* 1:31–40.

Wu SM, Maple BR (1998). Amino acid neurotransmitters in the retina: A functional overview. *Vision Res* 38:1371–1384.

Yamaguchi M, Hirai Y, Hage A *et al.* (1985) Small square (SS) net structure of the narrow Z-line. *J Mol Biol* 184:621–643 & 644.

Yanagida T (1985) Angle of active site of myosin heads in contractile muscle during sudden length changes. *J Muscle Res Cell Motil* 6:43–52.

Yule DI (2015) Ca^{2+} Signaling in Pancreatic Acinar Cells. Pancreapedia: Exocrine Pancreas Knowledge Base, doi:10.3998/panc.2015.24.

Zebelo SA, Matsui K, Ozawa R, Maffei ME (2012) Plasma membrane potential depolarisation and cytosolic calcium flux are early events involved in tomato (*Solanum lycopersicon*) plant-to-plant communication. *Plant Sci* 196:93–100.

Zima AV, Belevich AE, Povstyan AV, Kharkhun MI, Tsitsyura YaD, Shuba MF (1996) Mechanism of action of nitric oxide donors on voltage-activated calcium channels in vascular smooth muscle cells. *Neurophysiol* 28:232–239.

Index

absolute refractory period, 47
acetylcholine, 25, 33, 86, 95, 98, 110, 115, 129, 141, 147
acidification, 45
action potential, 22, 24, 26–33, 39–41, 43, 45–48, 51, 55, 56, 59, 96, 97, 102, 103, 107, 120–122, 125, 126, 129, 133, 139, 141, 145, 148, 151–154, 161, 162, 164, 165, 179, 180
active force, 36, 38, 39, 49, 50
adding electrons, 23
adenosine triphosphate, 13, 20, 44, 49, 83, 84, 178
adrenal, 16, 25, 106, 109
adrenal cortex, 16
adrenaline, 25, 109
adrenoreceptor, 25
aequorin, 37, 39, 51, 180
afferent, 33, 100, 113, 116, 119–125, 132, 134, 136
alkalinisation, 45
alternative theory, 71
amoebae, 168
ampere, 5
anaerobic, 150, 168, 173, 183
anionic phospholipids, 18, 45
anterior to posterior vector analysis, 157
antibiotic resistance, 174, 177
aqueous ionic medium, 11, 12
archaea, 173, 175–177

array of dipoles, 76, 79
arteriole, 86, 93, 95
atom, 1, 3–5, 7, 8, 14, 20, 44, 45, 48, 112, 114, 160
atropine, 100
autonomic nerve ending, 86

bacteria, 173–178, 182, 183
baroreceptor, 128, 129
baroreflex, 128, 130, 132, 133
bi-exponential decay, 60
bladder, 101, 102, 121
breathing, 116, 118, 132–135
brown fat cell, 16

C receptor, 119
Ca^{2+}, 6, 18, 33, 35–39, 41–46, 48–51, 53–60, 62, 64, 71, 72, 77, 80–82, 84, 85, 87, 89, 103–110, 112, 114, 125, 132, 136, 139, 145, 149, 162–164, 169, 172, 179, 180, 183
cable theory, 33
calcium ion, 6, 18, 35, 53, 60, 63, 83, 92, 139, 182
calcium overload, 35, 183
calcium-induced calcium release, 53, 55
cardiac conduction system, 152
cell membrane, 11, 12, 18, 35, 36, 39, 44, 85, 106, 110, 112, 120, 125, 126, 132, 135, 167, 169, 176, 177, 181, 183
chemoreceptor, 129, 131–134

chlorophyll, 158
chloroplast, 158, 159
cholecystokinin, 100
circulating Ca^{2+}, 50, 51
cold-tolerant species, 139, 140
conduction spread, 29
cyclic adenosine monophosphate, 25
cytoplasm, 11, 14, 15, 18, 20, 21, 23,
    25, 27, 29, 31, 34, 41, 44, 47, 48,
    53, 54, 56, 59, 80, 85, 97, 100, 104,
    107, 162, 163, 179

D-Luciferin, 179
decorated actin, 66, 67
delay of nearly 100 msec, 152
depolarisation, 22, 25, 28–33, 41, 42,
    44, 46, 47, 51, 52, 55, 56, 59, 60, 62,
    63, 93–95, 97, 98, 102, 106–110,
    113–115, 121, 125, 141, 145, 147,
    149, 151, 153, 154, 158, 161–165,
    167, 169, 170, 172, 179, 182, 183
detruser muscle, 102, 103
diastolic impedance, 25
diastolic interval, 22–25, 37
digestive tube, 97, 98
dihydropyridine receptor, 44
distribution of electrolytes, 11, 182
dopamine, 115
duct, 97, 99, 100, 105, 110
duration of this nerve impulse, 31
dyspnoea, 119

ear, 113, 136
efferent, 100, 124, 128, 129
electric birefringence, 79
electric eel, 2, 11
electric field force, 44, 45, 106
electric potential difference, 7, 18, 167
electrical restitution, 47
electrocardiograph, 3, 5, 151
electrode, 3, 12, 18, 29, 36, 39, 40,
    108, 118, 143, 170, 173
electrolysis, 12, 81
electromagnetic theory, 64, 66, 67, 73,
    74, 82
electron density, 14–18, 20–23, 25–27,
    29, 31–34, 36, 37, 39, 42, 44–48,

51, 53–55, 58, 59, 89, 90, 92–95,
    97, 98, 100, 102, 104, 106–113,
    120, 136, 139, 140, 145, 148, 153,
    158, 162–170, 172–176
electron flow, 1, 4, 11, 12, 14, 23, 25,
    27, 28, 31, 34, 40, 47, 48, 85, 90, 92,
    94, 95, 99, 100, 106, 107, 151, 153,
    158, 160–162, 164–167, 178, 183
electrostatic binding, 18
endocrine gland, 106, 110
endoplasmic reticulum, 35, 43, 54, 63
endothelial cell, 73, 90, 91, 93, 94
excitation-contraction coupling, 43,
    46, 179, 180
exocrine, 100, 110, 111, 136
external anal sphincter, 99
extracellular space, 15, 23
eye, 113, 114, 136, 141, 144, 172, 178

430 volts/sec, 28, 30, 31, 56
fat cell, 16
flow-mediated, 90–92
flow-mediated dilatation, 92
fluorescent Ca^{2+} indicator, 53, 56
frontal plane vector, 155
fungi, 167, 169, 175
funny current, 22, 24, 182

G protein, 25
gallbladder, 99, 100
gamma-aminobutyric acid, 115
gap junction, 28, 29, 91, 99, 100, 104,
    152
gel, 11, 12, 18, 54, 73, 79–81, 90, 93,
    143
glucose, 13, 94, 107, 169
glutamate, 115
glycocalyx, 90, 93, 94, 113

heart ventricular cell, 14
heavy meromyosin, 64, 66, 67, 69, 70,
    73, 75, 79–81, 179, 184
helical electrostatic field, 76, 82
His bundle and bundle branches, 152
Hodgkin and Huxley, 29
hydra, 171
hypoxia, 95, 131–133

impedance, 15, 23, 25–27, 31, 47, 107, 110, 120, 121, 125, 126, 136, 152, 154, 157, 163
insect, 66, 76, 77, 138–141
insect flight muscle, 66, 76, 77, 140
insulin, 107
intraluminal pressure, 86
invertebrate, 138–140, 145, 165
inward current, 6, 14, 22, 29, 46, 56, 92, 109, 143
ion flux balance, 36, 39, 40
Islet β cells of pancreas, 107
isometric test twitch, 50, 51

K^+, 6, 10–13, 17, 30, 32, 36, 39–42, 44, 82, 93, 99, 106, 108–110, 112, 114, 131, 132, 147, 169
$[K_o^+]/K_i^+]$, 16
K^+ ATP channels, 41
K^+ channels, 39–41
K^+ outflow, 40
kidney, 112, 113, 136

land plant, 149, 158, 161, 165, 170, 182
laser diffraction, 80
lateral instability, 66, 69, 71
leukocyte, 16
Leydig cell, 109
light-emitting organisms, 177
liver, 16, 99, 112, 136, 165
local $[Ca^{2+}]$, 71
longitudinal instability, 69
luminescence, 39, 178–180
lung, 95, 113, 116–119, 121, 131, 135, 136

M disc, 64, 65
mechanical restitution, 36, 37, 48, 50–52, 55, 60
microbiome, 175
micturition, 101
mitochondria, 13, 18–21, 25, 27, 31, 33, 34, 41, 42, 44, 48, 63, 85, 92, 98, 100, 104, 107, 121, 143, 150, 162, 164, 168, 173, 178, 179, 182, 183

mono-exponential decay, 60, 62
myelin sheath, 33

Na^+, 6, 10–14, 17, 29–32, 35–37, 39, 41, 42, 44, 46–48, 82, 98, 99, 107, 112, 114, 131, 149, 161, 165, 166, 169
Na^+ and K^+ distribution, 12
Na^+/Ca^{2+} exchanger, 35, 44
Na^+/K^+ ATPase, 6, 10, 13, 17, 36, 44, 112
NCX, 35, 36, 41, 44, 46, 48, 55, 59, 60, 104, 107, 108
Nernst equation, 7, 10
nerve, 3, 7, 16, 25, 31–33, 46, 86, 89, 95, 97–102, 106, 111, 113, 115, 116, 119–123, 128, 129, 133, 134, 136, 138, 142, 145
nerve axon spike, 31
neurotransmitter, 25, 33, 86, 95, 98, 100, 110, 114, 115, 129, 131, 132
neutron, 5
nitric oxide, 90, 91
NO synthase, 90, 92, 94
nodes of Ranvier, 33
non-myelinated, 33, 119, 120
noradrenaline, 25, 86, 89, 98, 115, 129
nucleus, 1, 5, 18, 168
nutrient recycling, 171

optimal contractile response, 49, 51, 62
organelle, 18, 20, 29, 43, 106, 110, 143, 178, 182
oscillation, 40, 114, 138–140, 170
outward current, 6, 23, 30
overlap, 66, 68, 69, 76, 77, 83, 121
Oxidative phosphorylation, 20
oxidative phosphorylation, 13, 20, 173, 182

pacemaker current, 27, 40, 102, 115, 182
pain receptor, 119
pancreatic acinar cell, 16, 110, 111
pancreatic islet cell, 16
papillary muscle, 46, 51, 60

parasympathetic, 25, 86, 95, 97–99, 101, 110, 128
parathyroid, 108
pathogen, 167, 177
pelvic parasympathetic nerve, 101
phages, 176, 177
phagocytosis, 168, 170
phosphatidylserine, 45
photosynthesis, 158, 160, 161, 163, 170, 171
pineal, 109
plankton, 170, 171, 176
plant intercommunication, 162
plateau, 27, 47, 164
platelet, 16, 115
post-paired pacing potentiation, 57
post-tachycardia potentiation, 51
post-voltage clamp potentiation, 57, 60
proton, 5, 20, 21, 45, 48, 113, 163, 164, 169, 173
pulsatility, 87, 88
Purkinje, 26, 27, 36, 37, 44, 104, 178

quantum, 1, 5, 68, 85, 140, 150, 181, 182, 184

recirculation fraction, 55, 57
red blood cell, 16, 17, 113, 181
relative refractory period, 47, 49
relaxation, 48–50, 53, 79, 88–90, 93, 99, 100
repolarisation, 25, 27, 31–34, 39–42, 48, 52–54, 62, 63, 93, 104, 153, 154, 162, 164, 165, 182
response to darkness, 163
retinal cell, 16
route through the heart of the electrical impulse, 152
ryanodine receptor, 44, 46, 53, 56, 109

salivary gland, 16, 100
salivary gland acinar cell, 16
sarcomere length, 37, 66, 68–70, 77, 78, 80, 82, 83
sarcoplasmic reticulum, 35, 43, 62, 87, 139
sea plant, 165

SERCA, 44, 49, 53, 54
serotonin, 95, 115
serotonin re-uptake mechanism, 115
sinus node cell, 16, 22–26, 39–41, 45, 89, 107
skeletal muscle, 16, 30, 33, 43, 44, 46, 49, 68, 77, 83, 86, 89, 99, 104, 116, 122, 124, 141, 147, 165
skinning, 36
sliding filament, 64
slime moulds, 169
Smooth muscle, 101, 102
smooth muscle, 86–89
sodium pump, 6, 10–13, 17, 36, 39–42, 44, 48, 82, 107, 108, 112, 113
sodium/calcium exchanger, 6
solenoid, 71–73
sphincter of Oddi, 99
splanchnic nerve, 100
stretch, 68–71, 75, 83, 87, 88, 98, 101, 116, 118, 121–126, 128, 132, 135, 138
stretch receptor, 121–123
stretch reflex, 123, 124
striated muscle, 37, 64, 69, 79, 86, 97, 101, 122, 125
sucrose gap, 51, 58
summarise our thoughts and predictions, 83
suspend thin filaments in space, 76, 82
sympathetic nervous system, 95
Systeme Internationale, 5

taste/smell receptor, 129
terminal cysternae, 53–56
test pulse interval, 49–52
thyroid follicular cell, 108
tone, 86–91, 95
touch receptor, 125, 126
trabeculae, 36, 37, 46
trains of impulses, 32
trans-membrane potential, 7–9, 11, 14, 16, 17, 22, 23, 26, 36, 39, 40, 42, 93, 181, 182
transient perturbation, 42
transverse tubule system, 44
troponin complex, 81, 82, 84

ureter, 101
urethra, 101, 102
urinary sphincter, 101

vagus nerve, 100, 116, 119, 123, 128
vascular endothelium, 16
vascular smooth muscle cell, 16, 92, 93
vasodilation, 90, 91, 93
vectorcardiogram, 154

vein, 86, 87, 91, 95
Velocity of shortening, 69, 70
Vernier effect, 65
viruses, 175–177

Weidmann, 26, 28, 31
wingbeat frequencies, 138

X-ray diffraction, 79, 80, 84

www.ingramcontent.com/pod-product-compliance
Lightning Source LLC
Chambersburg PA
CBHW050600190326
41458CB00007B/2121